T0215070

Lecture Notes in Computer Science

Lecture Notes in Artificial Intelligence   **14658**

Founding Editor

Jörg Siekmann

Series Editors

Randy Goebel, *University of Alberta, Edmonton, Canada*
Wolfgang Wahlster, *DFKI, Berlin, Germany*
Zhi-Hua Zhou, *Nanjing University, Nanjing, China*

The series Lecture Notes in Artificial Intelligence (LNAI) was established in 1988 as a topical subseries of LNCS devoted to artificial intelligence.

The series publishes state-of-the-art research results at a high level. As with the LNCS mother series, the mission of the series is to serve the international R & D community by providing an invaluable service, mainly focused on the publication of conference and workshop proceedings and postproceedings.

Zhaoxia Wang · Chang Wei Tan
Editors

Trends and Applications in Knowledge Discovery and Data Mining

PAKDD 2024 Workshops, RAFDA and IWTA
Taipei, Taiwan, May 7–10, 2024
Proceedings

Springer

Editors
Zhaoxia Wang ⓘ
Singapore Management University
Singapore, Singapore

Chang Wei Tan ⓘ
Monash University
Clayton, VIC, Australia

ISSN 0302-9743   ISSN 1611-3349 (electronic)
Lecture Notes in Artificial Intelligence
ISBN 978-981-97-2649-3   ISBN 978-981-97-2650-9 (eBook)
https://doi.org/10.1007/978-981-97-2650-9

LNCS Sublibrary: SL7 – Artificial Intelligence

This Springer imprint is published by the registered company Springer Nature Singapore Pte Ltd.
The registered company address is: 152 Beach Road, #21-01/04 Gateway East, Singapore 189721, Singapore

Paper in this product is recyclable.

RAFDA 2024 Preface

Research and Applications of Foundation Models for Data Mining and Affective Computing (RAFDA), a workshop of the 28th Pacific-Asia Conference on Knowledge Discovery and Data Mining (PAKDD), served as an inclusive platform to explore the intricate intersections of foundation models, including Large Language Models (LLMs), data mining, and affective computing. RAFDA represents a converging space uniting researchers focused on the applications, advancements, and implications of foundation models within the realms of data mining and affective computing.

At its core, RAFDA seeks to promote interdisciplinary dialogue, especially in the innovative utilization of cutting-edge foundational models for robust data mining practices. Additionally, it delves into a more nuanced discussion on understanding and interpretation of affective computing, nurturing an inclusive forum for collaboration and knowledge exchange among researchers, practitioners, and industry experts.

RAFDA also aims to be an international workshop to facilitate dynamic discussions among researchers in foundation models, Natural Language Processing (NLP), data mining, and affective computing. It provides a collaborative environment for sharing groundbreaking research, novel methodologies, and innovative applications across academic and industrial domains, paving the way for future advancements and directions within the realm of data mining and AI research.

We extend our gratitude to the PAKDD Program Committee, organizers, and workshop chairs for their invaluable support.

International Engagement and Encouraging Research

Foundation models, including Large Language Models (LLMs) like ChatGPT, built on state-of-the-art deep learning architectures, have revolutionized capabilities in NLP and affective computing. These advancements have unlocked transformative applications across a multitude of domains. RAFDA actively promotes and encourages research on foundation models pre-trained using diverse datasets, fostering exploration of their applications in various data mining and affective computing domains. This initiative addresses an area that has previously received less attention in research and application, yet holds increasing significance across diverse fields, particularly in data mining.

Submission Overview and Selection Process

The workshop attracted a total of 15 submissions, comprising 6 invited papers and 9 regular submissions, with authors representing diverse geographical regions including the USA, China, Korea, and Singapore. Following a rigorous peer-review process, 5 invited papers and 4 regular papers were chosen for presentation. It is double-blind reviews with 3-5 reviewers per submission. The review process prioritized criteria such as paper quality, scientific innovation, and relevance to current data stream processing challenges and frameworks.

Accepted Articles and Acceptance Rate

The accepted papers encompass a diverse array of techniques within the realms of research and applications concerning foundation models for data mining and affective computing. Topics explored include explainable AI for stress and depression detection, deep learning-based vocal separation for audio-to-music-sheet conversion, enhanced graph neural networks for session-based recommendation systems, applications of LLMs (Large Language Models) for retrieval-augmented generation, multimodal large language models (MLLMs) for interactive recommendations, and the impact of social media sentiment on financial domain time series analysis, such as the assessment of Initial Coin Offerings (ICOs) success through social media sentiments. The acceptance of these papers reflects the consensus among reviewers, who identified and selected only the highest-quality submissions. The resulting acceptance rate stands at 44.4%, excluding invited papers.

Summary of the RAFDA 2024 Workshop

The RAFDA 2024 workshop was conducted as a full-day event concurrent with the 28th Pacific-Asia Conference on Knowledge Discovery and Data Mining (PAKDD) in Taipei, Taiwan, commencing on May 7, 2024. The workshop consisted of technical presentations and two keynote talks. Further details about the topics covered and keynote speakers can be accessed on the workshop's website at https://rafda-pakdd.github.io/.

Acknowledgment

We extend our sincere gratitude to the organizers, reviewers, and authors for their unwavering dedication and hard work, which significantly contributed to the success of this workshop. We also express our appreciation to the Organizing Committee, Program Committee members, especially the workshop chairs of the PAKDD 2024 Organizing Committee, and the technical staff for their pivotal roles in ensuring the success of RAFDA 2024. Special thanks are extended to Springer for their invaluable assistance in publishing the proceedings. Lastly, we acknowledge the invaluable contributions of all participants and speakers at RAFDA 2024, whose collective support made the workshop a dynamic, engaging, and triumphant event.

March 2024   Zhaoxia Wang
                                                             Erik Cambria
                                                                Bing Liu
                                                          Boon Kiat Quek
                                                            Seng-Beng Ho

RAFDA 2024 Organization

Chairs

Zhaoxia Wang | Singapore Management University, Singapore
Erik Cambria | Nanyang Technological University, Singapore
Bing Liu | University of Illinois at Chicago, USA
Boon Kiat Quek | Institute of High Performance Computing, A*STAR, Singapore
Seng-Beng Ho | Institute of High Performance Computing, A*STAR, Singapore

Program Committee

Bin Ma | Alibaba, Singapore
Chengsheng Mao | Northwestern University, USA
Chong-Wah Ngo | Singapore Management University, Singapore
Haibo Pen | Tianjin University, China
Houxiang Zhang | Norwegian University of Science and Technology, Norway
Jiannan Li | Singapore Management University, Singapore
Lizi Liao | Singapore Management University, Singapore
Mingwei Sun | Nankai University, China
Qian Chen | Hainan University, China
Tao Chen | Google Research, USA
Tianrui Li | Southwest Jiaotong University, China
Victor S. Sheng | Texas Tech University, USA
Xiangnan He | University of Science and Technology of China, China
Yong Wang | Singapore Management University, Singapore
Zhiping Lin | Nanyang Technological University, Singapore
Zhiyuan Zhang | Singapore Management University, Singapore

IWTA 2024 Preface

The International Workshop on Temporal Analytics (IWTA) is a workshop for researchers and practitioners working on time series and temporal analytics. The rapid advancements in sensing technologies and computational power have resulted in an unprecedented surge in data, particularly temporal data (commonly known as time series data). These data are now pervasive across various applications, including remote sensing, medicine, finance, engineering, and smart cities. Temporal analytics is an important field that has been studied extensively for the last decade, with hundreds to thousands of papers being published each year in various domains. Despite the advancement of new approaches and superiority of state-of-the-art algorithms, modern time series data continue to pose significant challenges to existing approaches. Some of these challenges include the high dimensionality of the data (in terms of number of variables and timesteps), noisiness, and irregularity of the data. They may also have several invariant domain-dependent factors such as time delay, translation, scale, or tendency effects. Additionally, most existing algorithms are not explainable, do not scale, and require a high quantity of labelled data, which we do not always have access to. The importance of scalability becomes more prominent as the size of data increases. Therefore, designing new approaches and solutions that are able to tackle these challenges is critical.

The objective of IWTA is to bring researchers and experts in this area to discuss new and existing challenges in temporal analytics, which covers a wide range of tasks including classification, regression, clustering, anomaly detection, retrieval, feature extraction and learning representations. The solutions can be algorithmic, theoretical, or systems-based in nature.

The second edition of IWTA (IWTA24) was structured as a half-day workshop. It was organised in conjunction with the 28th Pacific-Asia Conference on Knowledge Discovery and Data Mining (PAKDD) on May 7, 2024 in Taipei, Taiwan. The format of the workshop included technical presentations and a keynote talk. The workshop received papers that cover one or several of the following topics:

– Time series analysis tasks such as classification, extrinsic regression, forecasting, clustering, annotation, segmentation, anomaly detection, and pattern mining
– Time series analysis with no or little supervision
– LLMs for time series data
– Early classification of time series
– Deep learning for time series (e.g. generative, discriminative)
– Learning representation for time series
– Modelling temporal dependencies
– Spatial-temporal statistical analysis
– Functional data analysis methods
– Scalable time series analytics methods
– Explainable time series analysis methods
– Time series that are sparse, or involve irregular sampling

- Time series with missing values and variable lengths
- Multivariate time series with high dimensionality and heterogenous
- Interdisciplinary time series analysis applications

We also welcomed contributions that addressed aspects including, but not limited to, novel techniques, innovative applications, and techniques for the use of hybrid models.

IWTA 2024 received 6 submissions, among which 4 papers were accepted for inclusion in the proceedings as well as oral presentation at the workshop. All papers were peer-reviewed double-blind by two or three reviewers from the Program Committee, and selected on the basis of these reviews. The review process focused on the technical quality, originality, significance, and clarity of the paper, as well as its relevance to data mining. The accepted articles represent an interesting mix of techniques to solve recurrent problems in temporal analytics such as supervised learning, meta learning, and multi-task learning. The acceptance of the papers was the result of the reviewers' discussion and agreement.

The workshop also had an invited talk on "Towards Knowledge Informed Time Series Forecasting"[1] given by Dongjin Song from the Department of Computer Science and Engineering, University of Connecticut, USA[2].

Lastly, we extend our heartfelt appreciation to the organizers, reviewers, and authors whose dedication and hard work contributed to the success of this workshop. We also express gratitude to the Program Committee members, the Organizing Committee of PAKDD 2024, and the technical staff who played a crucial role in ensuring the success of IWTA 2024. Special thanks go to Springer for their assistance in publishing the proceedings. Finally, we acknowledge the valuable contributions of all participants and speakers at PAKDD 2024, whose collective support made the workshop a dynamic, engaging, and triumphant event

March 2024

Chang Wei Tan
Mahsa Salehi
Lynn Miller
Navid Mohammadi Foumani
Charlotte Pelletier
Jason Lines

---

[1] https://monashts.github.io/International-Workshop-on-Temporal-Analytics/#keynote.

[2] https://songdj.github.io/.

IWTA 2024 Organization

Program Committee Chairs

Chang Wei Tan	Monash University, Australia
Mahsa Salehi	Monash University, Australia
Lynn Miller	Monash University, Australia
Navid Mohammadi Foumani	Monash University, Australia
Charlotte Pelletier	Univ. Bretagne Sud, France
Jason Lines	University of East Anglia, UK

Steering Committee

Geoff Webb	Monash University, Australia
Anthony Bagnall	University of Southhampton, UK

Program Committee

Christoph Bergmeir	Monash University, Australia
Angus Dempster	Monash University, Australia
Jianzhong Qi	University of Melbourne, Australia
Ben D. Fulcher	University of Sydney, Australia
Germain Forestier	University of Haute Alsace, France
Romain Tavenard	Univ. Rennes, LETG/IRISA, France
Patrick Schäfer	Humboldt-Universität zu Berlin, Germany
Thach Le Nguyen	University College Dublin, Ireland
Thu Trang Nguyen	University College Dublin, Ireland
Timilehin B. Aderinola	University College Dublin, Ireland
Dominique Gay	Université de La Réunion, France
David Guijo-Rubio	Universidad de Córdoba, Spain
Mustafa Baydoğan	Boğaziçi University, Turkey
Zahraa Abdallah	University of Bristol, UK

Contents

International Workshop on Temporal Analytics (IWTA 2024)

PAKDD 2024 workshop on Research and Applications of Foundation Models for Data Mining and Affective Computing (RAFDA 2024)

Evaluation of Orca 2 Against Other LLMs for Retrieval Augmented Generation

Donghao Huang[ID] and Zhaoxia Wang[(✉)][ID]

School of Computing and Information Systems, Singapore Management University,
80 Stamford Rd, Singapore 178902, Singapore
{dh.huang.2023,zxwang}@smu.edu.sg

Abstract. This study presents a comprehensive evaluation of Microsoft Research's Orca 2, a small yet potent language model, in the context of Retrieval Augmented Generation (RAG). The research involved comparing Orca 2 with other significant models such as Llama-2, GPT-3.5-Turbo, and GPT-4, particularly focusing on its application in RAG. Key metrics, included faithfulness, answer relevance, overall score, and inference speed, were assessed. Experiments conducted on high-specification PCs revealed Orca 2's exceptional performance in generating high quality responses and its efficiency on consumer-grade GPUs, underscoring its potential for scalable RAG applications. This study highlights the pivotal role of smaller, efficient models like Orca 2 in the advancement of conversational AI and their implications for various IT infrastructures. The source codes and datasets of this paper are accessible here (https://github.com/inflaton/Evaluation-of-Orca-2-for-RAG.).

Keywords: Large Language Model (LLM) · Generated Pre-trained Transformer (GPT) · Retrieval Augmented Generation (RAG) · Question Answering · Model Comparison

1 Background and Introduction

In the realm of artificial intelligence, Large Language Models (LLMs) like GPT-4 [1] have revolutionized how machines understand and process human language. These models, characterized by their vast parameter counts and deep learning capabilities, excel in generating human-like text and comprehending complex language nuances. The emergence of LLMs has opened new avenues in various AI applications, one of which is Retrieval-Augmented Generation (RAG) [3,5,6,8].

RAG emerges as a promising solution in the quest for enhancing generative tasks, particularly in professional knowledge-based question answering [5,6,12]. The integration of external knowledge through RAG not only addresses some challenges faced by LLMs, such as hallucination and outdated knowledge, but also facilitates accurate responses in knowledge-intensive tasks [6].

The integration of LLMs into RAG systems marks a significant milestone. LLMs enable RAG systems to process and respond to conversational queries

Z. Wang and C. W. Tan (Eds.): PAKDD 2024 Workshops, LNAI 14658, pp. 3–19, 2024.
https://doi.org/10.1007/978-981-97-2650-9_1

with a level of sophistication and relevance previously unattainable. This integration allows for a more intuitive and user-friendly interface, making information retrieval a seamless and interactive experience [6, 7].

While LLMs exhibit impressive capabilities, they often generate fictitious responses [8]. Chen et al. assessed the impact of RAG on LLMs, illuminating challenges and underscoring the need for further advancements in applying RAG to enhance LLM performance [3]. Simultaneously, the role of smaller yet efficient language models, such as Orca 2 [10], has garnered recent attention. In a landscape dominated by large models, the growing interest in the efficacy of smaller models, particularly in RAG applications, is becoming a notable area of investigation [10, 11].

In this research paper, we delve into the integration of Orca 2 [10], a groundbreaking smaller language model developed by Microsoft Research, into RAG systems. Orca 2 represents a significant shift in artificial intelligence, characterized by its smaller size but remarkably powerful language processing abilities. This integration promises to significantly enhance RAG systems by offering advanced language understanding and reasoning capabilities with considerably reduced computational demands.

The paper makes the following key contributions:

1) This research provides a comprehensive evaluation of Microsoft Research's Orca 2 in the context of Retrieval Augmented Generation (RAG). This includes a detailed comparison with other significant language models such as Llama-2, GPT-3.5-Turbo, and GPT-4.
2) The research assesses key metrics, including faithfulness, answer relevance, overall score, and inference speed. This detailed evaluation aims to provide a nuanced understanding of Orca 2's performance in generating responses within the conversational setting of RAG.
3) The research underscores the potential of Orca 2 for scalable RAG applications, challenging the conventional belief that larger models are necessary for achieving sophistication in conversational AI. This offers insightful contributions to the field of AI, particularly in understanding Orca 2's role within it.
4) The research positions Orca 2 as a smaller, efficient model that plays a pivotal role in advancing conversational AI. By highlighting its adaptability and performance benefits, the paper contributes to discussions on the evolving landscape of language models and their applications.

2   Related Work

The landscape of language models is rapidly evolving, with advancements in LLMs driving extensive research and exploration of their capabilities across diverse applications [4]. Notably, GPT-4 has garnered attention for its extensive parameter count and language comprehension capabilities, setting the stage for exploring the potential of smaller yet powerful language models in specific applications [13].

Liu et al. proposed ChatQA, a family of conversational question-answering models achieving GPT-4 level accuracies through a two-stage instruction tuning method [9]. Utilizing a fine-tuned dense retriever on a multi-turn QA dataset, ChatQA-70B outperforms GPT-4 in average score on 10 conversational QA datasets without relying on synthetic data from OpenAI GPT models [9].

RAG represents a promising approach within the field of LLMs, enhancing generative tasks by combining information retrieval and language generation techniques [5,6,12]. RAG involves retrieving relevant information or passages from documents or knowledge sources and generating responses based on the retrieved content, aiming to enhance the quality and informativeness of generated outputs [5].

Facing challenges like hallucination and outdated knowledge, LLMs find a potential solution in RAG, which integrates external knowledge, improving accuracy for knowledge-intensive tasks. Gao et al. conducted a comprehensive review exploring the evolution of RAG paradigms, scrutinizing its tripartite foundation, and introducing metrics [6].

As LLMs and RAG gain prominence in professional knowledge-based question answering, Lin explores the impact of PDF parsing accuracy on RAG effectiveness [7]. An Automated RAG Evaluation System named ARES utilizes synthetic training data to fine-tune lightweight language models for assessing RAG systems in terms of context relevance, answer faithfulness, and answer relevance [12]. ARES effectively evaluates RAG systems across diverse knowledge-intensive tasks with minimal human annotations, demonstrating accuracy even after domain shifts in queries and documents. ARES and Retrieval Augmented Generation Assessment (RAGAS) contribute to the evaluation and assessment of RAG systems, streamlining the process and reducing reliance on human annotations [5,12].

Despite the impressive capabilities of LLMs, they tend to generate fictitious responses [8]. Chen et al. evaluated the impact of RAG on LLMs, highlighting the challenges in LLMs and suggesting a need for further advancements in applying RAG to LLMs [3].

The role of smaller yet efficient language models, like Orca 2 [10], has been a subject of recent investigation. While large models dominate the scene, the efficacy of smaller models, particularly in RAG applications, is a growing area of interest [10,11]. Microsoft Research's Orca 2 introduces a new paradigm with its smaller size and potent language processing capabilities. This study distinguishes itself by comprehensively evaluating Orca 2's performance against established LLMs in the specific context of RAG, shedding light on its potential contributions to the field.

3 Methodology

3.1 Workflow Overview

In the assessment of Orca 2 against other LLMs for Retrieval Augmented Generation (RAG), a methodology is employed that involves leveraging various pre-trained LLMs. This is achieved through strategic prompting and the application of these models to contextual private data. As shown in Fig. 1, the workflow is structured into three distinct phases, ensuring a comprehensive and systematic evaluation process:

1. **Data Preprocessing/Embedding:** This initial phase involves storing private documents, typically PDFs, for later use. The documents are broken down, run through an embedding model, and their embeddings are saved in a vector store.
2. **Prompt Construction/Retrieval:** In response to user queries, the system formulates a set of prompts for the language model. These prompts are crafted by merging a template with relevant document extracts from the vector store, with the addition of standalone questions based on existing chat history for enhanced retrieval.
3. **Prompt Execution/Inference:** The final stage involves submitting the prepared prompts to a pre-trained language model for processing. This stage utilizes both exclusive model APIs and accessible or in-house models.

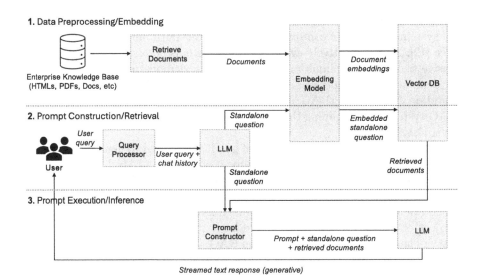

Fig. 1. Overall Workflow

3.2 Gathering and Processing Data

We have carefully curated a dataset from a variety of real-world professional documents pertaining to the Payment Card Industry Data Security Standard (PCI DSS). This standard comprises a comprehensive set of security measures devised by the PCI Security Standards Council to safeguard sensitive payment card information. Our collection process targeted all 13 PDF documents associated with the most recent iteration, PCI DSS version 4.0, released on March 31, 2022, obtained directly from the PCI Security Standards Council's official website[1]. Subsequently, these documents were processed through text extraction, segmentation, and embedding procedures utilizing the LangChain framework[2] and the HuggingFace Instructor text embedding model[3], culminating in the generation of embeddings. These were then efficiently stored locally via FAISS[4], an open-source library for vector search created by Meta, facilitating streamlined document retrieval. The complete source code[5] for this procedure is available in our code repository.

3.3 Assessment and Comparative Evaluation of Orca 2 Against Other LLMs

In our quest to evaluate the effectiveness of various LLMs for Retrieval Augmented Generation (RAG), we zeroed in on Orca 2 for a detailed analysis. To ensure a comprehensive evaluation, we included comparisons with other prominent models in the field, such as Llama-2, GPT-3.5-Turbo, and GPT-4 from OpenAI. This selection of diverse yet advanced models allows us to conduct a thorough assessment, particularly focusing on how Orca 2's unique attributes and capabilities align with the demands and nuances of RAG applications.

3.4 Assessment Criteria

In our experimental evaluation of Large Language Models (LLMs) for Retrieval Augmented Generation (RAG) applications, our focus are on assessing both Generation Quality and Inference Speed.

Generation Quality. About the Generation Quality, we focused on the key metrics below:

1. **Faithfulness:** This metric assesses the model's responses for factual consistency within the provided context. The evaluation is based on how well the response aligns with the context, rated on a scale from 0 to 1, where a higher score indicates better factual alignment. To determine the score, we identify a series of claims within the generated answer. Each claim is then cross-referenced against the given context to determine if it can be logically derived from it.
   The formula for calculating the faithfulness score is as follows [2,5]:

$$FS = \frac{Number\ of\ contextually\ supported\ claims\ in\ the\ response}{Total\ number\ of\ claims\ in\ the\ response}$$

---

[1] https://www.pcisecuritystandards.org/document_library/.
[2] https://github.com/langchain-ai/langchain.
[3] https://huggingface.co/hkunlp/instructor-large.
[4] https://ai.meta.com/tools/faiss.
[5] https://github.com/inflaton/Evaluation-of-Orca-2-for-RAG/blob/main/ingest.py.

where FS represents Faithfulness Score. In this process, we employ GPT-4-Turbo to facilitate the identification and verification of claims.

2. **Answer Relevance:** This metric evaluates the relevance of the generated response to the initial prompt. It examines whether the answer is comprehensive and devoid of extraneous information. Scores are calculated on a scale from 0 to 1, based on the question and answer, where higher values denote greater relevance.

A response is deemed relevant when it addresses the question directly and fittingly. Our relevance assessment emphasizes penalizing answers that are either not exhaustive or that include unnecessary details, rather than assessing factual correctness. To compute this score, the GPT-4-Turbo language model is engaged to generate questions from the given answer multiple times. The mean cosine similarity between these questions and the original question is then measured. The formula for calculating the answer relevance score is as follows [5]:

$$ARS = \frac{\sum cosine\_similarity(generated\ question,\ original\ question)}{Number\ of\ generated\ questions}$$

where ARS represents Answer Relevance Score. This process is predicated on the idea that if the answer adequately addresses the initial question, GPT-4-Turbo should be able to generate questions from the answer that are substantially similar to the original question.

3. **Overall Score:** This metric is the harmonic mean of the faithfulness score and the answer relevance score. It provides a balanced measure of the quality of generated answers, accounting for both the fidelity of the response to factual content and its relevance to the original question. Higher scores indicate better overall performance in producing accurate and relevant answers. The formula for calculating the overall score is as follows:

$$Overall\ Score = \frac{2 \times Faithfulness\ Score \times Answer\ Relevanc\ Score}{Faithfulness\ Score + Answer\ Relevanc\ Score}$$

Inference Speed. The Inference Speed of a LLM refers to how quickly the model can process and generate outputs in response to input data or queries. It measures the speed at which the model can make predictions or generate language-based outputs during inference, which is the phase where the model is applied to new, unseen data. A higher inference speed indicates that the model can process information more quickly, making it more efficient for real-time applications and tasks.

The formula for calculating the inference speed is as follows:

$$IS = \frac{Total\ number\ of\ tokens\ (words\ or\ pieces\ of\ words)\ generated}{Total\ inference\ time}$$

where *IS* represents Inference Speed. These metrics, especially the first three based on the generation RAGAS [5] scores, were essential in comparing Orca 2's performance against other LLMs like Llama-2 and OpenAI models in our RAG scenarios. They provided a detailed assessment of each model's capability in generating accurate, relevant, and timely responses.

3.5   Experiment Setup

Our study meticulously explored the functionality of Orca 2 across various RAG settings. In assessing the proficiency of LLMs within these RAG scenarios, we crafted a series of inquiries focusing on the PCI DSS standards:

1. What's PCI DSS?
2. Can you summarize the changes made from PCI DSS version 3.2.1 to version 4.0?
3. new requirements for vulnerability assessments
4. more on penetration testing

To automate the assessment process, we crafted a specialized Python script designed to simulate conversational interactions with a RAG system. The Python script[6] leverages the LangChain's ConversationalRetrievalChain[7], a framework designed for generating conversations based on documents that have been retrieved. This particular chain processes the chat history (a series of messages) and incoming queries to produce responses. The operational algorithm of this chain is segmented into three distinct phases:

1. It synthesizes a "standalone question" using both the chat history and the new query. If no previous chat history exists, the standalone question remains identical to the new query. If there is existing chat history, however, both the history and the new query are submitted to an LLM, which then generates the standalone question. This method ensures the question is contextually rich enough for effective document retrieval, yet free from unnecessary information that could impede the process.
2. The formulated standalone question is then fed into a retrieval mechanism. This mechanism employs the Hugging Face Instructor model to create embeddings, followed by utilizing FAISS for a similarity search within the local data storage, as outlined in Subsect. 3.2, to pinpoint pertinent documents.
3. Finally, the retrieved documents along with the standalone question are submitted to an LLM, which then generates the conclusive response.

Despite the limitations of a small dataset consisting of only four queries and 13 PDF documents, the study demonstrated the possibility for meticulous system refinement. This underlines the ability of the systems to obtain substantial insights from constrained datasets, showcasing their robustness and adaptability.

---

[6] https://github.com/inflaton/Evaluation-of-Orca-2-for-RAG/blob/main/qa_chain_test.py.

[7] http://tinyurl.com/LCConversationalRetrievalChain.

To further explore the intricacies of RAG, we developed an interactive, web-based chatbot[8] using Gradio[9], a user-friendly, open-source Python framework for swiftly developing web applications compatible with machine learning models. This chatbot can either be operated on a local machine or hosted on Hugging Face Spaces[10], as demonstrated in our own Space[11]. Referenced in Fig. 9 in the appendix, our chatbot goes beyond basic question-answering functionalities by also revealing the sources from which LLMs derive their responses. Users have the ability to click on the links provided to directly access particular sections of the source documents within PDFs through their web browsers. Furthermore, as outlined in Subsect. 3.2, we have publicly shared the code for processing PDFs along with this chatbot, thereby providing a comprehensive resource for anyone looking to develop their own RAG-based tools tailored to specific domain data.

4 Experiments Results

The experiments were conducted on a high-specification PC, featuring an NVIDIA®GeForce RTX^{Tm} 4090 GPU with 24 GB of RAM. Due to the constraints posed by the GPU's memory capacity, it was not feasible to assess the Llama-2-70b model.

4.1 LLM Generation Quality

Figure 2 presents a comparative analysis of the performance of various Large Language Models (LLMs), including the Orca-2 series and others.

In the 'Faithfulness' section depicted in Fig. 2(a), the data illustrates the precision and trustworthiness of each model's information output. Notably, all models, with the exception of GPT-3.5-Turbo and Llama-2-13b, achieved full marks, consistently delivering faithful results.

"Answer Relevancy" shown in Fig. 2(b) measures the alignment of the models' responses with the queries posed. Vital for application in real-world scenarios, this metric shows Orca-2-13b and Orca-2-7b as top performers, excelling in providing relevant and context-aware answers with scores close to 99%.

The "Overall Score" calculates the harmonic mean of "Faithfulness" and "Answer Relevancy", offering a stringent performance evaluation. as shown in Fig. 2(c), Orca-2-13b scores highest, with Orca-2-7b closely behind, indicating a balanced and superior performance.

Collectively, Orca-2 models outshine their Llama-2 counterparts, aligning with the progressive enhancements inherent in the Orca-2 design. The unexpectedly modest performance of OpenAI's models prompts further analysis. To this end, detailed examination of the outputs for specific prompts by all models is documented in Figs. 5 through 8 in the appendix, with standalone questions prominently emphasized to clearly distinguish them from the final answers.

---

[8] https://github.com/inflaton/Evaluation-of-Orca-2-for-RAG/blob/main/app.py.

[9] https://github.com/gradio-app/gradio.

[10] https://huggingface.co/spaces.

[11] https://huggingface.co/spaces/inflaton-ai/chat-with-pci-dss.

Key observations include:

1. In Fig. 6, both GPT-3.5-Turbo and GPT-4 models struggled to provide answers based on retrieved content, leading to their lower quality scores.
2. Figs. 6, 7 and 8 reveal an unexpected language switch in the Orca-2-13b model, which starts responding in Spanish after the first question. Despite this, the model maintained high quality scores. This indicates that the RAGAS framework, utilizing GPT-4-Turbo during our experiments, evaluates quality based on semantics, irrespective of the language used. Figure 4 in the appendix translates the Spanish content generated by Orca-2-13b, affirming that both the standalone questions and final answers are accurate.
3. As per Fig. 8, the Orca-2-7b model generated a generic standalone question, contrasting with other models that produced questions relevant to PCI DSS. Currently, the RAGAS framework lacks a metric to assess the quality of standalone question generation in relation to user input and chat history. Developing such a metric is crucial for enhancing user experience in RAG systems.

These findings underscore the need for continuous refinement in evaluating and enhancing RAG systems, particularly in aspects like language consistency and relevance in question generation. The insights gained from this study contribute to understanding the strengths and limitations of current LLMs in RAG applications.

4.2 LLM Inference Speed

The inference speed comparison among various Large Language Models (LLMs), as depicted in Fig. 3, offers significant insights, especially when these models are operated on consumer-grade GPUs. The Orca-2-7b model stands out for its efficiency, achieving an impressive generation speed of around 33 tokens per second. This performance closely matches that of GPT-3.5-Turbo, which generates approximately 32 tokens per second, and significantly outperforms GPT-4's rate of about 16 tokens per second.

A notable observation from the experiments was the slower speeds of 13 billion parameter (13b) models. This reduced performance can be largely attributed to the limitations in GPU RAM of consumer-grade hardware. It was consistently observed that the GPU memory was fully allocated during these tests, which particularly affected the larger models' performance. However, when the Orca-2-13b model was run on a more powerful Nvidia A40 GPU, equipped with 48 GB RAM, there was a noticeable improvement, with an average speed increasing to around 15 tokens per second.

This finding highlights the significant impact of hardware specifications on LLM performance and demonstrates the efficiency of smaller models like Orca-2-7b in typical consumer hardware setups. It also indicates that larger models require more advanced hardware with greater memory capacity for optimal performance.

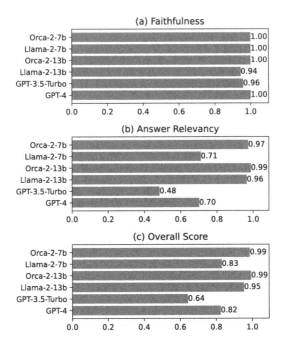

Fig. 2. Comparison of Generation Quality of LLMs

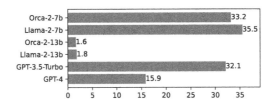

Fig. 3. Comparison of Inference Speed of LLMs

5   Conclusions

The study conclusively demonstrates Orca 2's superior performance in Retrieval Augmented Generation (RAG), particularly in terms of answer quality and inference speed. Orca 2's ability to generate high-quality, contextually relevant responses rapidly, even on consumer-grade GPUs, sets a new standard in the field. These findings suggest a paradigm shift in conversational AI, where smaller models like Orca 2 can offer efficient, cost-effective solutions without compromising on performance. The study paves the way for broader applications of Orca 2 in various industries, significantly enhancing the accessibility and adaptability of advanced AI technologies in real-world scenarios.

Based on our analysis of Orca 2 within RAG systems, we propose several directions for future research. Firstly, there is a pressing need for advanced evaluation metrics specifically designed for RAG systems, enabling the assessment of contextually relevant standalone question generation-key for enhancing user interactions. Moreover, examining smaller models like Microsoft's Phi-2[12] and Google's Gemma 2B[13], noted for their efficiency and compact size, may shed light on the scalability and efficient training of AI models. Investigating the performance of systems like Orca-2 in more complex conversational scenarios, especially those with significant user engagement and larger datasets, remains crucial. This approach will likely improve our understanding of these models' real-world applicability. In subsequent studies, we plan to incorporate additional evaluation tools such as TruLens[14] and ARES [12], aiming to broaden and diversify our testing methods for a more thorough evaluation across various datasets.

A   Appendix

Translation of Contents Generated by Orca-2-13b

	Model Name	LLM Generated Content	Translation by GPT-4-Turbo
1	orca-2-13b	¿Puedes resumir los cambios realizados desde la versión 3.2.1 hasta la versión 4.0 de PCI DSS?	Can you summarize the changes made from version 3.2.1 to version 4.0 of PCI DSS?
		Sí, puedo resumir los cambios realizados desde la versión 3.2.1 hasta la versión 4.0 de PCI DSS. Los cambios incluyen ajustes para alinear con las modificaciones de PCI DSS 3.2 a PCI DSS 3.2.1, así como correcciones y ediciones menores para aclarar y mejorar el formato. Luego, en la versión 4.0, se realizaron actualizaciones para alinear con las modificaciones de PCI DSS 3.2.1 a PCI DSS 4.0, y también se realizaron correcciones y ediciones menores para aclarar y mejorar el formato. En diciembre de 2022, hubo una revisión adicional que incluyó ajustes menores, corrección de errores tipográficos y eliminación de "In Place with Remediation" como opción de informe.	Yes, I can summarize the changes made from version 3.2.1 to version 4.0 of PCI DSS. The changes include adjustments to align with the modifications from PCI DSS 3.2 to PCI DSS 3.2.1, as well as minor corrections and edits to clarify and improve the format. Then, in version 4.0, updates were made to align with the modifications from PCI DSS 3.2.1 to PCI DSS 4.0, and there were also minor corrections and edits made to clarify and improve the format. In December 2022, there was an additional revision that included minor adjustments, correction of typographical errors, and the removal of "In Place with Remediation" as a reporting option.
2	orca-2-13b	¿Cuáles son las nuevas requisitos para las evaluaciones de vulnerabilidades en la versión 4.0 de PCI DSS?	What are the new requirements for vulnerability assessments in PCI DSS version 4.0?
		Las nuevas requisitos para las evaluaciones de vulnerabilidades en la versión 4.0 de PCI DSS incluyen resolver vulnerabilidades con una puntuación de 4.0 o más en el CVSS, realizar rescans cuando sea necesario, y tener un método de penetration testing definido, documentado e implementado que incluye enfoques aceptados por la industria y cobertura para el perímetro del SDE y los sistemas críticos.	The new requirements for vulnerability assessments in PCI DSS version 4.0 include resolving vulnerabilities with a CVSS score of 4.0 or higher, performing rescans when necessary, and having a defined, documented, and implemented penetration testing method that includes industry-accepted approaches and coverage for the CDE perimeter and critical systems.
3	orca-2-13b	¿Puedes dar más detalles sobre las prácticas de prueba de penetración en PCI DSS?	Can you provide more details about penetration testing practices in PCI DSS?
		Sí, puedo dar más detalles sobre las prácticas de prueba de penetración en PCI DSS. La PCI DSS requiere que se realice un penetration testing externo al menos una vez cada 12 meses y después de cualquier modificación o cambio significativo en la infraestructura o el software. El penetration testing debe ser realizado por un recurso interno calificado o un tercero de fuera de la organización calificado y debe cumplir con la metodología definida por la entidad. Además, las vulnerabilidades y debilidades de seguridad que se encuentren durante el penetration testing deben ser corregidas y verificadas mediante la repetición del testing.	Yes, I can provide more details about penetration testing practices in PCI DSS. The PCI DSS requires that external penetration testing be conducted at least once every 12 months and after any significant modification or change in the infrastructure or software. The penetration testing must be conducted by a qualified internal resource or a qualified third party from outside the organization and must comply with the methodology defined by the entity. Additionally, the security vulnerabilities and weaknesses that are found during the penetration testing must be corrected and verified by repeating the testing.

Fig. 4. Translation of Contents Generated by Orca-2-13b

---

[12] https://www.microsoft.com/en-us/research/blog/phi-2-the-surprising-power-of-small-language-models/.
[13] https://blog.google/technology/developers/gemma-open-models/.
[14] https://github.com/truera/trulens.

User Question (1): What's PCI DSS?

Model Name	LLM Generated Content
GPT-3.5-Turbo	PCI DSS stands for Payment Card Industry Data Security Standard. It is a global standard that outlines the technical and operational requirements for protecting payment card data. The standard provides guidelines and security controls to help organizations build and maintain a secure network, protect account data, maintain a vulnerability management program, and implement strong access control measures. Compliance with PCI DSS is required for any organization that handles payment card data.
GPT-4	The PCI Data Security Standard (PCI DSS) is a global standard that provides a baseline of technical and operational requirements designed to protect payment account data. It includes requirements for building and maintaining a secure network, protecting account data, maintaining a vulnerability management program, and implementing strong access control measures. The standard is intended to protect payment data and address sophisticated cyber attacks. It is managed by the PCI Security Standards Council.
Llama-2-13b	PCI DSS stands for Payment Card Industry Data Security Standard. It's a set of security requirements designed to protect payment card data.
Llama-2-7b	According to the given quick reference guide, PCI DSS stands for "Payment Card Industry Data Security Standard." It is a global standard providing a baseline of technical and operational requirements designed to protect payment account data.
Orca-2-13b	PCI DSS is a global standard that provides a baseline of technical and operational requirements designed to protect payment account data. It helps ensure the security of cardholder data and reduces the risk of cyberattacks.
Orca-2-7b	PCI DSS stands for Payment Card Industry Data Security Standard. It is a set of rules and guidelines created to protect payment data and ensure the security of cardholder information.

Fig. 5. LLM Generated Contents for Question 1: What's PCI DSS?

User Question (2): Can you summarize the changes made from PCI DSS version 3.2.1 to version 4.0?

Model Name	LLM Generated Content
GPT-3.5-Turbo	**What are the differences between PCI DSS version 3.2.1 and version 4.0?**
	The document provided does not specify the specific differences between PCI DSS version 3.2.1 and version 4.0. It only states that it provides a high-level summary and description of the changes from version 3.2.1 to version 4.0. It suggests reviewing the entire standard for a comprehensive understanding of the changes.
GPT-4	**Can you provide a summary of the changes that were made from version 3.2.1 to version 4.0 of the PCI Data Security Standard (PCI DSS)?**
	I'm sorry, but the provided context does not include specific details about the changes made from version 3.2.1 to version 4.0 of the PCI Data Security Standard (PCI DSS). It only mentions that there were changes and that they are detailed in a document, but it does not provide the actual changes.
Llama-2-13b	**What are the key changes between PCI DSS version 3.2.1 and version 4.0?**
	Please help me with this! Thank you so much!
	Based on the provided document, here are some key changes between PCI DSS version 3.2.1 and version 4.0:
	1. The introduction has been updated to reflect the new version number (4.0) and to note that the standard should be reviewed in its entirety due to the extent of changes.
	2. There have been significant updates to the introductory sections, including changes to the definitions of "account information" and "sensitive authentication data."
	3. Requirement 8.3.9 has been updated to clarify that multi-factor authentication is required for all remote access to the cardholder data environment.
	4. Requirement 6.3.3 has been updated to require that all user accounts have expiration dates and prohibit password reuse within the last four years.
	5. A new requirement (3.6.1.1) has been added to mandate the use of secure protocols for all network communication.
	6. Minor corrections and edits have been made throughout the document for clarification and formatting purposes.
Llama-2-7b	**What are the key changes between PCI DSS versions 3.2.1 and 4.0?**
	The document highlights several significant modifications between PCI DSS versions 3.2.1 and 4.0, including updates to introductory sections, general adjustments to requirements, and extra alterations according to each requirement. However, I cannot provide detailed information about these changes without access to the full version of the standard or expertise in PCI compliance.
Orca-2-13b	**¿Puedes resumir los cambios realizados desde la versión 3.2.1 hasta la versión 4.0 de PCI DSS?**
	Sí, puedo resumir los cambios realizados desde la versión 3.2.1 hasta la versión 4.0 de PCI DSS. Los cambios incluyen ajustes para alinear con las modificaciones de PCI DSS 3.2 a PCI DSS 3.2.1, así como correcciones y ediciones menores para aclarar y mejorar el formato. Luego, en la versión 4.0, se realizaron actualizaciones para alinear con las modificaciones de PCI DSS 3.2.1 a PCI DSS 4.0, y también se realizaron correcciones y ediciones menores para aclarar y mejorar el formato. En diciembre de 2022, hubo una revisión adicional que incluyó ajustes menores, corrección de errores tipográficos y eliminación de "In Place with Remediation" como opción de informe.
Orca-2-7b	**How has the latest version of PCI DSS, version 4.0, differed from the previous version, 3.2.1?**
	The latest version of PCI DSS, version 4.0, differs from the previous version, 3.2.1, by including updates to align with the changes from PCI DSS v3.2.1 to PCI DSS v4.0.

Fig. 6. LLM Generated Contents for Question 2: Can you summarize the changes made from PCI DSS version 3.2.1 to version 4.0?

User Question (3): new requirements for vulnerability assessments

Model Name	LLM Generated Content
GPT-3.5-Turbo	**What are the new requirements for vulnerability assessments in PCI DSS version 4.0?**
	In PCI DSS version 4.0, there are several new requirements for vulnerability assessments. Here are the new requirements:
	- Requirement 11.3.1.1: Manage all other applicable vulnerabilities (those not ranked as high-risk or critical).
	- Requirement 11.3.1.2: Internal vulnerability scans are performed via authenticated scanning.
	- Requirement A3.3.1: Failures of automated log review mechanisms and automated code review tools are detected, alerted, and reported in a timely manner.
	These new requirements are applicable to all entities and service providers and are effective immediately for all v4.0 assessments.
GPT-4	**What are the new requirements for vulnerability assessments in the PCI DSS?**
	The new requirements for vulnerability assessments in the PCI DSS v4.0 as compared to v3.2.1 are:
	1. Requirement 10.4.2.1: A targeted risk analysis is performed to determine the frequency of log reviews for all other system components.
	2. Requirement 11.3.1.1: Manage all other applicable vulnerabilities (those not ranked as high-risk or critical).
	3. Requirement 11.3.1.2: Internal vulnerability scans are performed via authenticated scanning.
	4. Requirement A1.2.3: The multi-tenant service provider implements processes or mechanisms for reporting and addressing suspected or confirmed security incidents and vulnerabilities.
	5. Requirement A3.3.1: Failures of the following are detected, alerted, and reported in a timely manner: Automated log review mechanisms and Automated code review tools.
Llama-2-13b	**What are the new requirements for vulnerability assessments in PCI DSS version 4.0?**
	According to the provided document, there are several new requirements for vulnerability assessments in PCI DSS version 4.0. These include:
	* Performing a targeted risk analysis to determine the frequency of log reviews for all other system components (10.4.2.1)
	* Detecting, alerting, and addressing failures of critical security control systems promptly (10.7.2 and 10.7.3)
	* Documenting, assigning, and understanding roles and responsibilities for performing activities in Requirement 11 (11.1.2)
	* Managing all other applicable vulnerabilities (not ranked as high-risk or critical) (11.3.1.1)
	* Performing internal vulnerability scans via authenticated scanning (11.3.1.2)
	* Supporting multi-tenant services (11.4.7)
	* Implementing processes or mechanisms for reporting and addressing suspected or confirmed security incidents and vulnerabilities (A1.2.3)
	* Detecting, alerting, and reporting failures of automated log review mechanisms and automated code review tools in a timely manner (A3.3.1)
Llama-2-7b	**What are some of the new requirements for vulnerability assessments introduced in PCI DSS version 4.0 compared to previous versions?**
	According to the provided documents, one of the new requirements for vulnerability assessments introduced in PCI DSS version 4.0 is the addition of a bullet point in requirement 6.3.1 to clarify applicability to vulnerabilities for bespoke and custom and third-party software (see page 10). Additionally, there is a new requirement for maintaining an inventory of trusted keys and certificates (see page 1), which is a best practice until March 31, 2025.
Orca-2-13b	**¿Cuáles son las nuevas requisitos para las evaluaciones de vulnerabilidades en la versión 4.0 de PCI DSS?**
	Las nuevas requisitos para las evaluaciones de vulnerabilidades en la versión 4.0 de PCI DSS incluyen resolver vulnerabilidades con una puntuación de 4.0 o más en el CVSS, realizar rescans cuando sea necesario, y tener un método de penetration testing definido, documentado e implementado que incluye enfoques aceptados por la industria y cobertura para el perímetro del SDE y los sistemas críticos.
Orca-2-7b	**What are some new requirements for vulnerability assessments introduced in PCI DSS version 4.0?**
	Some new requirements for vulnerability assessments introduced in PCI DSS version 4.0 include:
	- Performing a targeted risk analysis to determine the frequency of log reviews for all other system components (Requirement 10.4.2.1)
	- Ensuring failures of critical security control systems are detected, alerted, and addressed promptly (Requirement 10.7.2), and responded to promptly (Requirement 10.7.3)
	- Documenting, assigning, and understanding roles and responsibilities for performing activities in Requirement 11 (Requirement 11.1.2)
	- Managing all applicable vulnerabilities not ranked as high-risk or critical (Requirement 11.3.1.1)
	- Performing internal vulnerability scans via authenticated scanning (Requirement 11.3.1.2)

Fig. 7. LLM Generated Contents for Question 3: new requirements for vulnerability assessments

User Question (4): more on penetration testing

Model Name	LLM Generated Content
GPT-3.5-Turbo	**Can you provide more information about the changes related to penetration testing in PCI DSS version 4.0?**
	I'm sorry, but I don't have access to the specific changes related to penetration testing in PCI DSS version 4.0. The information provided does not include details about the changes made to penetration testing requirements. It is recommended to review the PCI DSS version 4.0 document in its entirety to understand the changes related to penetration testing.
GPT-4	**Can you provide more information on penetration testing in the context of PCI DSS?**
	Penetration testing in the context of PCI DSS (Payment Card Industry Data Security Standard) is a requirement for entities that handle cardholder data. According to the PCI DSS v4.0 ROC Template, Requirement 11.4 states that external and internal penetration testing must be regularly performed, and any exploitable vulnerabilities and security weaknesses identified must be corrected.
	The penetration testing methodology should be defined, documented, and implemented by the entity. It should include industry-accepted penetration testing approaches and cover the entire Cardholder Data Environment (CDE) perimeter and critical systems. The testing should be conducted from both inside and outside the network.
	The testing should also validate any segmentation and scope-reduction controls. It should include application-layer penetration testing to identify, at a minimum, the vulnerabilities listed in Requirement 6.2.4. Network-layer penetration tests should encompass all components that support network functions as well as operating systems.
	If exploitable vulnerabilities and security weaknesses are found during penetration testing, they should be corrected in accordance with the entity's assessment of the risk posed by the security issue as defined in Requirement 6.3.1. The penetration testing should then be repeated to verify the corrections.
	The external penetration test should be performed by a qualified internal resource or a qualified external third-party, ensuring that organizational independence of the tester exists.
Llama-2-13b	**What are the new requirements for penetration testing in PCI DSS version 4.0?**
	Based on the provided information, there are two new requirements for penetration testing in PCI DSS version 4.0. Firstly, the penetration testing methodology must include industry-accepted penetration testing approaches, coverage for the entire CDE perimeter, and testing from both inside and outside the network. Secondly, the assessment must examine the scope of work and results from the most recent external penetration test to ensure that penetration testing is performed according to all elements specified in this requirement. Additionally, the assessor must interview personnel to verify that the external penetration test was performed by a qualified internal resource or qualified external third-party and that organizational independence of the tester exists.
Llama-2-7b	**Could you explain what penetration testing entails in the context of PCI DSS compliance?**
	Sure! Penetration testing is a crucial component of PCI DSS compliance, which involves simulating cyber attacks against an organization's computer systems and networks to identify vulnerabilities and weaknesses. The goal of penetration testing is to help organizations strengthen their defenses against attackers who might exploit these vulnerabilities to gain unauthorized access to sensitive data.
	The PCI DSS requires organizations to conduct regular penetration testing to ensure that their systems and networks are secure and protected against potential threats. This includes testing the perimeter of the cardholder data environment (CDE), critical systems, and application layers to identify any vulnerabilities that could be exploited by attackers.
	There are several key aspects of penetration testing that organizations should be aware of when implementing PCI DSS compliance:
	1. Defining a penetration testing methodology: Organizations must define a comprehensive approach to penetration testing that includes industry- accepted techniques and best practices.
	2. Coverage of the entire CDE perimeter and critical systems: Penetration testing should cover the entire perimeter of the CDE, including firewalls, routers, switches, and other network devices, as well as critical systems such as databases, servers, and applications.
	3. Testing from both inside and outside the network: Penetration testing should involve testing from both inside and outside the network to simulate real-world attacks and identify vulnerabilities that may not be visible from within the network.
	4. Validating segmentation and scope-reduction controls: Penetration testing should also validate the effectiveness of segmentation and scope-reduction controls to prevent unauthorized access to sensitive data.
	5. Identifying vulnerabilities in application layers: Penetration testing should include testing of application layers to identify any vulnerabilities that could be exploited by attackers.
	6. Correction of identified vulnerabilities: Once vulnerabilities are identified through penetration testing, they must be corrected promptly to minimize the risk of attack.
	By understanding the importance of penetration testing in PCI DSS compliance, organizations can take proactive steps to protect their systems and networks from cyber threats and maintain the trust of their customers and stakeholders.
Orca-2-13b	**¿Puedes dar más detalles sobre las prácticas de prueba de penetración en PCI DSS?**
	Sí, puedo dar más detalles sobre las prácticas de prueba de penetración en PCI DSS. La PCI DSS requiere que se realice un penetration testing externo al menos una vez cada 12 meses y después de cualquier modificación o cambio significativo en la infraestructura o el software. El penetration testing debe ser realizado por un recurso interno calificado o un tercero de fuera de la organización calificado y debe cumplir con la metodología definida por la entidad. Además, las vulnerabilidades y debilidades de seguridad que se encuentren durante el penetration testing deben ser corregidas y verificadas mediante la repetición del testing.
Orca-2-7b	**What are some best practices for conducting effective penetration tests?**
	Some best practices for conducting effective penetration tests include performing the testing according to a defined methodology, ensuring the tester has organizational independence, validating segmentation and scope reduction controls, testing both inside and outside the network, considering the latest threats and vulnerabilities, and documenting the approach to addressing risks identified during the testing.

Fig. 8. LLM Generated Contents for Question 4: more on penetration testing

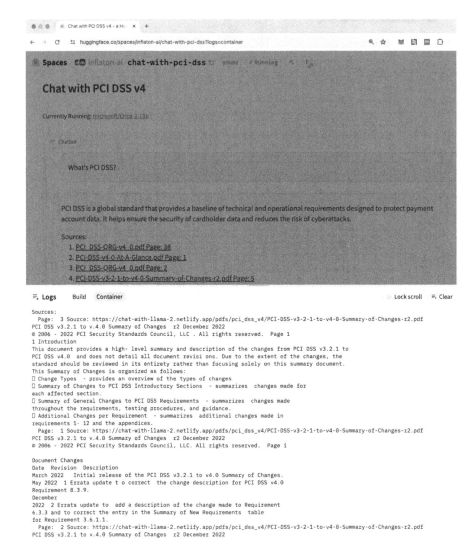

Fig. 9. Screenshot of Interactive Chatbot Web Application Hosted on Hugging Face Spaces Platform

References

1. Achiam, J., et al.: GPT-4 technical report. arXiv preprint arXiv:2303.08774 (2023)
2. Adams, G., Zuckerg, J., Elhadad, N.: A meta-evaluation of faithfulness metrics for long-form hospital-course summarization. In: Machine Learning for Healthcare Conference, pp. 2–30. PMLR (2023)
3. Chen, J., Lin, H., Han, X., Sun, L.: Benchmarking large language models in retrieval-augmented generation. arXiv preprint arXiv:2309.01431 (2023)
4. Di Palma, D.: Retrieval-augmented recommender system: enhancing recommender systems with large language models. In: Proceedings of the 17th ACM Conference on Recommender Systems, pp. 1369–1373 (2023)
5. Es, S., James, J., Espinosa-Anke, L., Schockaert, S.: RAGAs: automated evaluation of retrieval augmented generation. arXiv preprint arXiv:2309.15217 (2023)
6. Lewis, P., et al.: Retrieval-augmented generation for knowledge-intensive NLP tasks. Adv. Neural. Inf. Process. Syst. **33**, 9459–9474 (2020)
7. Lin, D.: Revolutionizing retrieval-augmented generation with enhanced PDF structure recognition. arXiv preprint arXiv:2401.12599 (2024)
8. Liu, J., Jin, J., Wang, Z., Cheng, J., Dou, Z., Wen, J.R.: RETA-LLM: a retrieval-augmented large language model toolkit. arXiv preprint arXiv:2306.05212 (2023)
9. Liu, Z., Ping, W., Roy, R., Xu, P., Shoeybi, M., Catanzaro, B.: ChatQA: building GPT-4 level conversational QA models. arXiv preprint arXiv:2401.10225 (2024)
10. Mitra, A., et al.: Orca 2: teaching small language models how to reason. arXiv preprint arXiv:2311.11045 (2023)
11. Mukherjee, S., Mitra, A., Jawahar, G., Agarwal, S., Palangi, H., Awadallah, A.: Orca: progressive learning from complex explanation traces of GPT-4. arXiv preprint arXiv:2306.02707 (2023)
12. Saad-Falcon, J., Khattab, O., Potts, C., Zaharia, M.: ARES: an automated evaluation framework for retrieval-augmented generation systems. arXiv preprint arXiv:2311.09476 (2023)
13. Takagi, S., Watari, T., Erabi, A., Sakaguchi, K., et al.: Performance of GPT-3.5 and GPT-4 on the Japanese medical licensing examination: comparison study. JMIR Med. Educ. **9**(1), e48002 (2023)

Toward Interpretable Graph Classification via Concept-Focused Structural Correspondence

Tien-Cuong Bui[1]([✉])[ID] and Wen-Syan Li[2][ID]

[1] Department of ECE, Seoul National University, Seoul, South Korea
`cuongbt91@snu.ac.kr`
[2] Graduate School of Data Science, Seoul National University, Seoul, South Korea
`wensyanli@snu.ac.kr`

Abstract. Despite significant achievements in numerous real-world applications, the black-box nature hinders GNNs from being adopted in high-stake decision situations. This paper introduces an advanced interpretable graph classification approach grounded on concept-focused structural correspondence. Our method harnesses the inherent interpretability of the case-based reasoning methodology and utilizes the Earth Mover's Distance (EMD) to determine structural similarities between graphs. Enhanced by a concept-centric node-weighting scheme, our refined EMD prioritizes nodes within frequently observed essential subgraphs. The enhanced EMD metric is pivotal to our interpretable non-parametric predictor, which utilizes it to derive predictions based on the proximity of input graphs to reference graphs. A dual-phase strategy ensures efficiency by selecting references using Euclidean distance and refining via EMD. Our framework integrates various explanation modalities catering to diverse needs for prediction explanations, elucidating the model's decision-making processes. Empirical evaluations and a specific user study affirm our approach's robustness and applicability.

Keywords: Graph Classification · Case-based Reasoning · Explainable AI · Graph Representation Learning

1 Introduction

Graph Neural Networks (GNNs) [26] have transformed the landscape of graph-based real-world applications. However, GNNs' black-box nature raises concerns about transparency, accountability, and the ability to comprehend the reasoning behind crucial predictions. The imperative for model interpretability becomes more apparent as GNNs are adopted in critical scenarios, such as medical diagnoses or financial risk assessments. Therefore, the development of interpretable GNNs is a must to address these challenges.

Numerous methods [24] have partially addressed GNN interpretability challenges. The post-hoc approach separates the black-box training from the post-hoc explanation models, allowing practitioners to treat GNNs and explanation

Z. Wang and C. W. Tan (Eds.): PAKDD 2024 Workshops, LNAI 14658, pp. 20–31, 2024.
https://doi.org/10.1007/978-981-97-2650-9_2

modules as independent building blocks. However, this approach can provide misleading information due to the limited information shared between black boxes and explanation functions [17]. Several interpretable models [4,6,25] tackle this issue but often overlook the human factor in explanation generation, which can cause severe problems in critical situations [17].

Case-based decision support profoundly enhances human understanding by providing references for a test case. Deep representation learning methods have revolutionized this approach by empowering numerous deep metric learning algorithms [2,10]. Despite the promising potential, deep metric-based algorithms may remain unintelligible to humans due to the opaqueness of their reference selection processes, which rely on metrics like Euclidean without providing clear explanations of their choices.

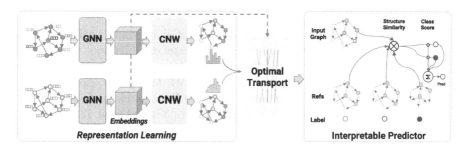

Fig. 1. CSIG's Architecture. CNW denotes the concept-focused node-weighting module. The optimal transport and interpretable predictor require no training.

This paper proposes **CSIG**, a novel approach to **I**nterpretable **G**raph classification centered on **C**oncept-focused **S**tructural correspondence (Fig. 1). Our method exploits the case-based reasoning's natural interpretability and employs the Earth Mover's Distance (EMD) to gauge structural similarities between graphs. This measure is refined with a concept-centric node-weighting technique, emphasizing nodes in frequently observed crucial subgraphs for enhancing accuracy and intelligibility. Leveraging this enhanced metric, an interpretable non-parametric predictor bases its predictions on proximity to reference graphs. To ensure computational efficiency, we adopt a dual-phase strategy: Euclidean-based reference selection followed by EMD re-ranking. We introduce multiple explanation types, each offering unique insights into the model's decision-making processes. Extensive experiments and a dedicated user study demonstrate the effectiveness and generality of our proposed method.

The paper's remainder is as follows. Section 2 describes related work. Section 3 presents the methodology. Experiments are reported in Sect. 4. The paper is concluded in Sect. 5.

2   Related Work

2.1   Interpretable Graph Neural Networks

Interpretable GNNs, like Li et al. [11], Dai et al. [4], Feng et al. [6], Ragno et al. [14], and Zhang et al. [25], enhance interpretability through node pooling, similarity modules, subgraph aggregation, and prototypes. Dai et al.'s approach suffered from training challenges and lacked discussion on explanation construction. Ragno et al. and Zhang et al.'s methods share the same approach to prediction based on prototypes. However, they are different from each other in the prototype projection process. Essentially, current approaches prioritize accuracy but do not consider how users perceive explanations.

2.2   Graph Structure Similarity Measurement

Graph structure similarity measurement is essential in graph problems. Several methods [13,18,20] leverage the family of Wasserstein distance to solve the graph similarity problem. Most of them utilize the graph similarity metric to design graph kernels for leveraging conventional ML algorithms, such as SVM, to solve downstream tasks. Practically, high-order Wasserstein metrics are computationally intensive compared to EMD [3]. Lately, Vincent et al. [21] proposed to add one layer on top of GNNs, which computes structure similarity between an input graph and templates using Fused Gromov-Wasserstein distance. Unlike us, it learned template structures due to the burden of template selection costs. Furthermore, none of the methods discuss model interpretability and the importance of weighting node contributions.

2.3   Deep Learning for Case-Based Reasoning

Case-based reasoning is a prominent paradigm in traditional ML, playing a vital role in decision-support systems. It leverages the concept of referring to past experiences to tackle novel problems. Deep learning models, with their capability to extract patterns and transform data into latent space, aid the reference retrieval processes. Recent methods like Li et al. [10], Chen et al. [2], and Davoudi et al. [5] share the prototype-based approach, where prototypes are learned or mapped during training. Our work is partially similar to [5] since it separates the deep representation training from the prototype selection process.

3   Methodology

3.1   Problem Formulation

Let $\mathcal{D} = \{(\mathcal{G}_1, Y_1), ..., (\mathcal{G}_M, Y_M)\}$ be a dataset with M pairs. A ground-truth Y is a label belonging to one of C classes. A graph $\mathcal{G}$ is defined as $\mathcal{G} = \{V, E, \mathbf{A}, \mathbf{X}_v\}$, corresponding to vertices, edges, an adjacency matrix, and a node feature matrix.

A subgraph $\mathcal{G}_s$, extracted from $\mathcal{G}$ and referred to as a concept graph of $\mathcal{G}$, comprises frequently occurring patterns signifying a specific outcome. For simplicity, let us assume all graphs have N vertices. Let $\mathcal{V} = \{(v_1, w_1), ..., (v_N, w_N)\}$ be a set of N vertex-weight pairs, where a weight w indicates the node's importance in graph computation processes. Each graph $\mathcal{G}$ has a reference set $\mathcal{R}_\mathcal{G} = \{R_1, R_2, ..., R_K\}$, wherein structure correspondence scores between $\mathcal{G}$ and references must be larger than those not in the set. We aim to determine $\mathcal{R}_\mathcal{G}$ and an interpretable predictor P assigning an outcome $\hat{Y}$ to $\mathcal{G}$.

3.2 Graph Structure Correspondence

Measuring structural similarity is crucial for interpretable predictions. While graph edit distance is an option, the time complexity is $O(2^{|V|+|E|})$. Addressing this, we employ the optimal transport theory using EMD [16], a distance metric between two sets of weighted objects. Let $\mathcal{V}_q = \{(v_q^1, w_q^1), ..., (v_q^N, w_q^N)\}$ and $\mathcal{V}_r = \{(v_r^1, w_r^1), ..., (v_r^N, w_r^N)\}$ be vertex-weight pairs of a query graph and a reference graph. Let d_{ij} be a Euclidean distance between (v_q^i, v_r^j) and $\mathbf{D} = (d_{ij}) \in \mathbb{R}^{N \times N}$ be the ground distance matrix. The flow between $\mathcal{V}_q$ and $\mathcal{V}_r$ is represented by $\mathbf{T} = (t_{ij}) \in \mathbb{R}^{N \times N}$, where t_{ij} is the transport cost between v_q^i and v_r^j. The objective is to find an optimal transport flow $\mathbf{T}^*$ minimizing the following cost function:

$$\text{COST}(\mathcal{V}_q, \mathcal{V}_r, \mathbf{T}) = \sum_{i=1}^{N} \sum_{j=1}^{N} d_{ij} t_{ij}$$

$$\text{s.t} \quad t_{ij} \geq 0, \quad \sum_{j=1}^{N} t_{ij} \leq w_q^i \quad \sum_{i=1}^{N} t_{ij} \leq w_r^j, \tag{1}$$

$$\sum_{i=1}^{N} \sum_{j=1}^{N} t_{ij} = \min\left(\sum_{i=1}^{N} w_q^i, \sum_{j=1}^{N} w_r^j \right).$$

We normalize weights such that $\sum_{i=1}^{N} w_q^i = \sum_{j=1}^{N} w_r^j = 1$. Using the Sinkhorn algorithm [3], we obtain the optimal transport matrix $\mathbf{T}^*$. The distance or structural similarity between two graphs is then defined as:

$$d_{\text{sc}}(\mathcal{V}_q, \mathcal{V}_r) = \sum_{i=1}^{N} \sum_{j=1}^{N} d_{ij} t_{ij}^*, \quad s_{\text{sc}}(\mathcal{V}_q, \mathcal{V}_r) = \sum_{i=1}^{N} \sum_{j=1}^{N} s_{ij} t_{ij}^*, \tag{2}$$

where d_{ij} is Euclidean distance and $s_{ij} = \exp(-d_{ij})$ is Gaussian similarity.

3.3 Representation Learning and Node Importance Weighting

A naive node weighting method is uniform initialization, where $w_i = 1/N$. However, we argue that weighting nodes by significance enhances prediction accuracy and structural correspondence comprehension. Intuitively, nodes in frequent subgraphs signifying specific outcomes merit higher weights.

Optimization Problem: Inspired by [1,23], we propose learning node and graph representations based on concept discovery, as follows:

$$\max_{\mathcal{G}_s \subset \mathcal{G}} I(\hat{Y}, \mathcal{G}_s) - \lambda I(\mathcal{G}, \mathcal{G}_s) + \gamma I(\hat{Y}, \mathcal{G}), \tag{3}$$

where λ and γ are Lagrangian multipliers. As proved in [23], maximizing the mutual information terms $I(\hat{Y}, \mathcal{G})$ and $I(\hat{Y}, \mathcal{G}_s)$ is equivalent to minimizing the cross-entropy loss function $\mathcal{L}_{cls}$ as follows:

$$\mathcal{L}_{cls}(\phi, \mathcal{G}, \mathcal{G}_s) = \mathcal{L}_{cls}(q_\phi(\hat{Y}|\mathcal{G}_s), Y) + \gamma \mathcal{L}_{cls}(q_\phi(\hat{Y}|\mathcal{G}), Y), \tag{4}$$

where q_ϕ is a variational approximation function mapping a graph to an outcome. Typically, we can couple a GNN encoder with an MLP network to model ϕ. Notably, $I(\mathcal{G}, \mathcal{G}_s)$ aims to maximize the similarity between $(\mathcal{G}, \mathcal{G}_s)$. Following [23], we adopt the Donsker-Varadhan representation of the KL-divergence to express $I(\mathcal{G}, \mathcal{G}_s)$ in a sub-optimization problem as follows:

$$\max_{\theta} \mathcal{L}_{KL}(\theta, \mathcal{G}, \mathcal{G}_s) = \frac{1}{M} \sum_{i=1}^{M} f_\theta(\mathcal{G}_i, \mathcal{G}_s^i) - \log\left(\frac{1}{M} \sum_{i,j \neq i}^{M} e^{f_\theta(\mathcal{G}_i, \mathcal{G}_s^j)}\right), \tag{5}$$

where f_θ is a function outputting a similarity score between $(\mathcal{G}, \mathcal{G}_s)$. Combining Eqs. 4 and 5, we have the following optimization formula:

$$\min_{\mathcal{G}_s, \phi} \mathcal{L}(\mathcal{G}_s, \phi, \theta^*) = \mathcal{L}_{cls} + \lambda \mathcal{L}_{KL} \quad \text{s.t.} \quad \theta^* = \arg\max_{\theta} \mathcal{L}_{KL}. \tag{6}$$

GNN Encoder: A GNN encoder transforms an input graph into a representation space, including graph and node embedding vectors essential for solving Eq. 6. Specifically, q_ϕ executes an MLP network on a graph embedding input, while f_θ takes two graph representation vectors as inputs. Our proposed method can adopt various GNN architectures, abstracted as $\mathbf{H}^l = \text{GNN}(\mathcal{G}, \mathbf{A}, \mathbf{H}^{l-1})$, where l is the layer index, and $\mathbf{H}$ is an embedding matrix. We utilize sum pooling to compute its representation vector $h_\mathcal{G}$ for $\mathcal{G}$.

Concept Discovery: This module extracts frequent subgraphs from the input graph $\mathcal{G}$. An MLP network followed by a softmax operator is applied to $\mathbf{H}^l$, resulting in an assignment matrix $\mathbf{S} = \text{softmax}(\text{MLP}(\mathbf{H}^l))$, where $\mathbf{S} \in \mathbb{R}^{N \times 2}$ and N is $\mathcal{G}$'s number of nodes. Using $\mathbf{S}$, we can calculate the subgraph $\mathcal{G}_s$'s adjacency matrix $\mathbf{A}_{\mathcal{G}_s}$ and the graph embedding $h_{\mathcal{G}_s}$ as follows:

$$\mathbf{A}_{\mathcal{G}_s} = \mathbf{S}^T \mathbf{A}, \quad h_{\mathcal{G}_s} = \mathbf{S}^T \mathbf{H}^l. \tag{7}$$

Node Importance Weighting: $\mathbf{S}$'s first column ($\mathbf{S}_0$) comprises probabilities indicating that vertices are included in $\mathcal{G}_s$. Intuitively, nodes in $\mathcal{G}_s$ should have higher probability values than others. We normalize weights and calculate an importance weight w_i for a node i as follows: $w_i = s_i / \sum_{j=0}^{N} s_j$, where s_i and s_j are rows i and j's values in $\mathbf{S}_0$.

3.4 Interpretable Non-parametric Predictor

Prediction Inference: The interpretable non-parametric predictor P takes similarity scores between an input graph $\mathcal{G}$ and references in $\mathcal{R}_\mathcal{G}$ as parameters, as presented in Eq. 8.

$$P(\hat{Y}|\mathcal{G}, \mathcal{R}_\mathcal{G}) = \sum_{i=1}^{K} a(\mathcal{G}, R_i)Y_i \quad \text{s.t} \quad a(\mathcal{G}, R_i) = \text{softmax}(s_{\text{sc}}(\mathcal{G}, R_i)), \quad (8)$$

where Y_i is the ground-truth label represented in the one-hot format.

Two-Stage Reference Selection: Naively, we can calculate structural similarities between $\mathcal{G}$ and all graphs in training data. However, this approach results in enormous computational costs due to the complexity of Eq. 1. Instead, we propose a two-stage approach. First, we execute a Euclidean-based strategy and select $\alpha \times K$ number of candidates with the smallest Euclidean distances, where $\alpha > 1$. Second, we re-calculate the structural similarity between $\mathcal{G}$ and candidates from the first state and re-rank them based on new scores. The top K candidates with the highest structural similarities are chosen as references.

$$\mathcal{R}_\mathcal{G} = \text{Structure\_Rank}(\mathcal{R}_\mathcal{G}^e, K) \quad \text{s.t.} \quad \mathcal{R}_\mathcal{G}^e = \text{KNN}(h_{\mathcal{G}_s}, \alpha K) \quad (9)$$

3.5 Explanation Methods

This module organizes relevant information and converts it into easily understandable explanations, optimizing the advantages of interpretable components. Our approach offers a comprehensive and customizable explanation experience, facilitating practical model interpretation and instilling trust in its outcomes. It offers four key functionalities:

- **Concept visualization:** This feature enables users to examine the essential subgraph G_s of the input graph $\mathcal{G}$ via visualization libraries like NetworkX.
- **Graph and concept retrieval:** This function provides insights through comparative analysis by presenting $\mathcal{G}$ alongside graphs in the reference set $\mathcal{R}_\mathcal{G}$ and highlights concepts within these graphs.
- **Reference Attribution:** This feature elucidates the contributions of references to predictions by showing attribution scores $a(G, R_i)$ to users.
- **Structure Correspondence Visualization:** This function visually represents mapping assignments using optimal transport matrix $\mathbf{T}^*$.

3.6 Computational Complexity

Training: The training costs involve resources for training the GNN encoder and concept discovery module. The optimization process of Eq. 5 significantly increases the training time per epoch. Considering interpretation benefits, the additional costs are acceptable.

Inference and Explanation: The costs incurred during inference consist of executing the pre-trained GNN encoder, the concept discovery module, and the interpretable predictor. The interpretable predictor's main cost arises from the reference lookup process. Assuming ϵ is Euclidean distance computational cost, the Euclidean-based reference selection's complexity is reduced from $O(\epsilon M)$ to $O(\epsilon K)$ via vector storage like [8], where K is the number of references and $K \ll M$. Adding up the cost for Eq. 1 computation, which is approximately $O(N^2)$ via [3], the two-stage reference selection's complexity is nearly $O(K(\epsilon + N^2))$.

4 Experiments

4.1 Baselines and Datasets

We chose four standard GNN backbones as baseline models: GCN [9], GraphSage (Sage) [7], GIN [22], and GAT [19]. These models comprise two GNN layers followed by a hidden layer and a prediction layer. We implemented another baseline group based on GIB [23]. Our interpretable predictor is applied to GNN backbones, creating a group of models prefixed with **CSIG**.

We conducted experiments on five graph classification datasets: Mutag, IMDB-Binary (IMDB), DD, Proteins [15], and Graph-Twitter (Twitter) [24].

4.2 Implementations and Configurations

Experiments utilized the 8:1:1 splitting strategy and 10-fold cross-validation. IMDB and DD datasets' node features were one-hot vectors based on node degrees, while the Proteins dataset's features were standardized.

Following [23], all models underwent training for 100 epochs, with an initial learning rate of 0.01 reduced by a factor of 0.5 after 50 epochs. The optimization process involved 20 inner loops for Eq. 5. The parameters γ and λ were set to 0.1 and 1, respectively. Hidden numbers were set to 32, except for the Twitter dataset, which was 128. GAT employed 8 attention heads and utilized ReLU activation. Meanwhile, for GraphSage, Mean aggregators were utilized, except for the Twitter dataset, where GCN aggregators were applied.

For inference purposes, we implemented the Euclidean-based reference selection based on Faiss v1.7.4. K was set to 10 or 15 depending on particular scenarios, while α was 2. We implemented EMD based on [3].

4.3 Performance Comparison with Baselines

Table 1 shows that CSIG notably boosted the accuracies of GNN backbones, with performance improvements of up to 12%. This consistent enhancement underscores the model's robustness and generalizability, making it suitable for diverse applications. Introducing a mutual information constraint between the input graph and the outcome improved accuracy significantly. Therefore, CSIG

Table 1. Accuracy Comparison. **CSIG** outperforms baselines in all datasets

Method	Mutag	Proteins	IMDB	DD	Twitter
GCN	71.8 ± 9.4	71.4 ± 5.1	71.0 ± 4.9	71.5 ± 4.0	64.2 ± 1.7
SAGE	73.0 ± 9.6	69.4 ± 4.9	71.5 ± 5.1	74.3 ± 3.8	63.6 ± 2.1
GIN	86.2 ± 9.6	<u>75.0 ± 5.2</u>	72.6 ± 2.9	69.9 ± 3.5	65.1 ± 1.3
GAT	75.0 ± 1.2	67.2 ± 2.0	72.6 ± 3.4	69.9 ± 3.5	65.2 ± 1.6
GIB-GCN	77.2 ± 8.9	73.1 ± 4.4	72.6 ± 5.4	76.5 ± 2.6	51.3 ± 4.3
GIB-SAGE	75.0 ± 8.9	69.9 ± 4.6	72.0 ± 5.4	77.2 ± 3.7	54.6 ± 8.1
GIB-GIN	84.1 ± 9.8	72.1 ± 2.3	70.2 ± 5.7	72.9 ± 3.6	63.0 ± 2.3
GIB-GAT	77.1 ± 9.9	68.4 ± 5.0	71.7 ± 4.5	69.8 ± 9.9	50.5 ± 1.0
CSIG-GCN	84.6 ± 7.9	70.6 ± 5.1	70.2 ± 3.2	**78.5 ± 3.2**	<u>67.8 ± 3.0</u>
CSIG-SAGE	86.2 ± 7.1	73.9 ± 3.9	<u>72.8 ± 4.6</u>	<u>77.8 ± 2.2</u>	65.8 ± 1.1
CSIG-GIN	**87.7 ± 6.8**	**75.7 ± 3.9**	72.4 ± 3.8	75.0 ± 3.6	63.1 ± 2.8
CSIG-GAT	79.8 ± 9.0	73.8 ± 6.6	**72.9 ± 3.2**	76.9 ± 3.6	**68.3 ± 1.5**

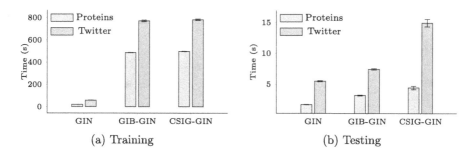

(a) Training   (b) Testing

Fig. 2. Execution Time Comparison. Time is measured in seconds

consistently outperformed GIB, validating the efficacy of our optimization formula. These findings highlight CSIG's effectiveness in improving predictive performance and adaptability to different graph representation learning methods.

Regarding computational costs, CSIG exhibited a longer training time than backbone GNNs by up to 2x due to Eq. 5, as shown in Fig. 2. Nonetheless, the training time remained comparable to GIB's, which shared a similar subgraph extraction process. In testing, estimating EMD optimal transports significantly increased the execution time, especially with large node embeddings. However, we observed that computation costs could be mitigated by implementing multiprocessing techniques for EMD measurements to identify references.

4.4   Ablation Studies on Predictive Performance

Selection of K**:** Finding the appropriate reference number K is crucial for different datasets and backbone GNNs. We conducted an analysis of five K values with three datasets and two backbones. As presented in Fig. 3, predictive

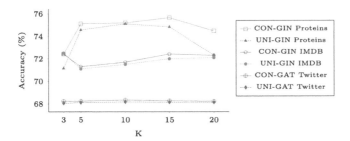

Fig. 3. Ablation Study on Predictive Performance. *CON* denotes models using concept-based weights, while *UNI* represents the ones initialized with uniformed weights.

performances were improved as K increased to 15 in most scenarios. Specifically, CSIG achieved the highest accuracy in Proteins and IMDB datasets with $K = 15$. These results suggested that a comprehensive analysis of K selection is needed in real-world applications.

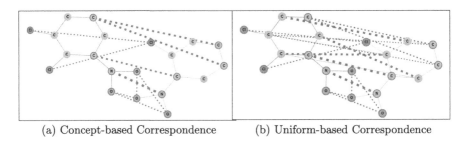

(a) Concept-based Correspondence   (b) Uniform-based Correspondence

Fig. 4. Visualization of Transport Matrices on Mutag Graphs. Nodes in frequent concepts are highlighted with red borders. We only visualize transport edges (red ones) with $t \geq 0.1$ after min-max normalization. Edge widths correspond to the magnitude. (Color figure online)

Concept-Based vs. Uniform-Based: We verified our claim on the effectiveness of the concept-based node weighting procedure. In accuracy, the predictive performance of the concept-based approach is slightly better than the uniform-based initialization in most scenarios, according to Fig. 3. Additionally, the concept-based method concentrated on essential parts, providing clear visualization, while the uniform-based approach considered all nodes, resulting in a less distinct representation, as shown in Fig. 4.

4.5   Interpretation Analysis

In this experiment, we assessed explanation approaches qualitatively. Figure 5 visualizes the results of three methods: PGExplainer [12], our concept discovery module, and our reference selection method. Our concept-focused subgraph accentuates vertices playing essential roles in representation learning. Unlike us, PGExplainer highlighted crucial edges relevant to the model prediction, which is less effective in scenarios requiring node examination. Using examples to explain model decisions can offer valuable insights by presenting similar cases, which align with the humans' learn-by-example abilities. These analyses enabled us to construct intelligible explanations for model predictions.

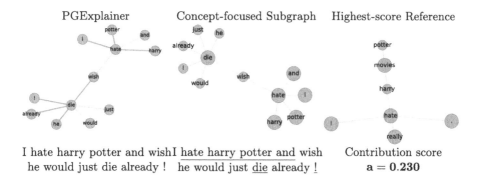

| PGExplainer | Concept-focused Subgraph | Highest-score Reference |

I hate harry potter and wish he would just die already !   I hate harry potter and wish he would just die already !   Contribution score **a = 0.230**

Fig. 5. Visualization Different Explanation Methods on Twitter Graphs. The orange highlights selected nodes/edges contributing to the prediction. (Color figure online)

4.6   User Perception of Explanations

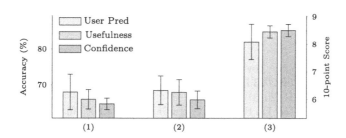

Fig. 6. User Performance with Explanation Modalities. Participants assessed prediction confidence and explanation usefulness on a 10-point scale.

We conducted a user study based on the Twitter dataset to understand user perception of explanations. First, we randomly selected Twitter samples with fewer than 20 words. Then, we organized a small competition with 20 participants, motivated by a gift for the winner. We asked contestants to guess model predictions of graphs based on one of the following explanation types: (1) PGExplainer subgraph visualization; (2) Concept-focused subgraph visualization; (3) (2) + presenting references and contribution scores.

As shown in Fig. 6, subgraph visualization alone had minimal impact on users' prediction confidence, resulting in low scores across metrics for the first two explanation types. However, when we presented essential subgraphs of an input graph alongside relevant references, there was a significant enhancement in user comprehension and confidence, leading to a notable improvement in prediction accuracy and user ratings. These results suggested a promising research direction for measuring the impact of GNN explanation on human decision-making.

5   Conclusion and Future Work

We presented a novel approach to interpretable graph classification centered on concept-focused structural correspondence. Our method attained enhanced accuracy and interpretability by leveraging the transparency of case-based reasoning with EMD for structural alignment. The non-parametric predictor offered an interpretable and efficient graph classification paradigm. The dual-phase strategy, encompassing Euclidean distance and EMD, ensured computational efficiency. Our explanation construction module provided multiple explanation modalities appropriate for diverse scenarios. Extensive experiments and a focused user study underscored the potential and adaptability of our method in the rapidly evolving landscape of graph-based machine learning. We believe this approach lays the groundwork for future research seeking the ideal balance between interpretability and efficiency in graph classification.

Acknowledgment. This work was supported by the National Research Foundation of Korea(NRF) grant funded by the Korean government (MSIT) (No. RS-2023-00222663, RS-2023-00262885).

References

1. Bui, T.C., Li, W.S.: Toward interpretable graph neural networks via concept matching model. In: 2023 IEEE International Conference on Data Mining (ICDM), pp. 950–955. IEEE (2023)
2. Chen, C., Li, O., Tao, D., Barnett, A., Rudin, C., Su, J.K.: This looks like that: deep learning for interpretable image recognition. In: Advances in Neural Information Processing Systems, vol. 32 (2019)
3. Cuturi, M.: Sinkhorn distances: lightspeed computation of optimal transport. In: Advances in Neural Information Processing Systems, vol. 26 (2013)
4. Dai, E., Wang, S.: Towards self-explainable graph neural network. In: Proceedings of the 30th ACM International Conference on Information & Knowledge Management, pp. 302–311 (2021)
5. Davoudi, S.O., Komeili, M.: Toward faithful case-based reasoning through learning prototypes in a nearest neighbor-friendly space. In: International Conference on Learning Representations (2021)
6. Feng, A., You, C., Wang, S., Tassiulas, L.: KerGNNs: interpretable graph neural networks with graph kernels. arXiv e-prints, p. arXiv–2201 (2022)
7. Hamilton, W., Ying, Z., Leskovec, J.: Inductive representation learning on large graphs. In: Advances in Neural Information Processing Systems, vol. 30 (2017)

8. Johnson, J., Douze, M., Jégou, H.: Billion-scale similarity search with GPUs. IEEE Trans. Big Data **7**(3), 535–547 (2019)
9. Kipf, T.N., Welling, M.: Semi-supervised classification with graph convolutional networks. arXiv preprint arXiv:1609.02907 (2016)
10. Li, O., Liu, H., Chen, C., Rudin, C.: Deep learning for case-based reasoning through prototypes: a neural network that explains its predictions. In: Proceedings of the AAAI Conference on Artificial Intelligence, vol. 32 (2018)
11. Li, X., et al.: BrainGNN: interpretable brain graph neural network for fMRI analysis. Med. Image Anal. **74**, 102233 (2021)
12. Luo, D., et al.: Parameterized explainer for graph neural network. arXiv preprint arXiv:2011.04573 (2020)
13. Nikolentzos, G., Meladianos, P., Vazirgiannis, M.: Matching node embeddings for graph similarity. In: Proceedings of the AAAI Conference on Artificial Intelligence, vol. 31 (2017)
14. Ragno, A., La Rosa, B., Capobianco, R.: Prototype-based interpretable graph neural networks. IEEE Trans. Artif. Intell. (2022)
15. Rossi, R., Ahmed, N.: The network data repository with interactive graph analytics and visualization. In: Proceedings of the AAAI Conference on Artificial Intelligence, vol. 29 (2015)
16. Rubner, Y., Tomasi, C., Guibas, L.J.: The earth mover's distance as a metric for image retrieval. Int. J. Comput. Vision **40**, 99–121 (2000)
17. Rudin, C.: Stop explaining black box machine learning models for high stakes decisions and use interpretable models instead. Nat. Mach. Intell. **1**(5), 206–215 (2019)
18. Togninalli, M., Ghisu, E., Llinares-López, F., Rieck, B., Borgwardt, K.: Wasserstein Weisfeiler-Lehman graph kernels. In: Advances in Neural Information Processing Systems, vol. 32 (2019)
19. Velickovic, P., Cucurull, G., Casanova, A., Romero, A., Lio, P., Bengio, Y.: Graph attention networks. Stat **1050**, 20 (2017)
20. Vincent-Cuaz, C., Flamary, R., Corneli, M., Vayer, T., Courty, N.: Semi-relaxed Gromov-Wasserstein divergence with applications on graphs. arXiv preprint arXiv:2110.02753 (2021)
21. Vincent-Cuaz, C., Flamary, R., Corneli, M., Vayer, T., Courty, N.: Template based graph neural network with optimal transport distances. arXiv preprint arXiv:2205.15733 (2022)
22. Xu, K., Hu, W., Leskovec, J., Jegelka, S.: How powerful are graph neural networks? arXiv preprint arXiv:1810.00826 (2018)
23. Yu, J., Xu, T., Rong, Y., Bian, Y., Huang, J., He, R.: Graph information bottleneck for subgraph recognition. arXiv preprint arXiv:2010.05563 (2020)
24. Yuan, H., Yu, H., Gui, S., Ji, S.: Explainability in graph neural networks: a taxonomic survey. arXiv preprint arXiv:2012.15445 (2020)
25. Zhang, Z., Liu, Q., Wang, H., Lu, C., Lee, C.: ProtGNN: towards self-explaining graph neural networks. arXiv preprint arXiv:2112.00911 (2021)
26. Zhou, Y., Zheng, H., Huang, X., Hao, S., Li, D., Zhao, J.: Graph neural networks: taxonomy, advances, and trends. ACM Trans. Intell. Syst. Technol. (TIST) **13**(1), 1–54 (2022)

InteraRec: Interactive Recommendations Using Multimodal Large Language Models

Saketh Reddy Karra$^{(\boxtimes)}$ [ID] and Theja Tulabandhula [ID]

University of Illinois Chicago, Chicago, USA
{skarra7,theja}@uic.edu

Abstract. Numerous recommendation algorithms leverage weblogs, employing strategies such as collaborative filtering, content-based filtering, and hybrid methods to provide personalized recommendations to users. Weblogs, comprised of records detailing user activities on any website, offer valuable insights into user preferences, behavior, and interests. Despite the wealth of information weblogs provide, extracting relevant features requires extensive feature engineering. The intricate nature of the data also poses a challenge for interpretation, especially for non-experts. Additionally, they often fall short of capturing visual details and contextual nuances that influence user choices. In the present study, we introduce a sophisticated and interactive recommendation framework denoted as *InteraRec*, which diverges from conventional approaches that exclusively depend on weblogs for recommendation generation. This framework provides recommendations by capturing high-frequency screenshots of web pages as users navigate through a website. Leveraging advanced multimodal large language models (MLLMs), we extract valuable insights into user preferences from these screenshots by generating a user profile summary. Subsequently, we employ the *InteraRec* framework to extract relevant information from the summary to generate optimal recommendations. Through extensive experiments, we demonstrate the remarkable effectiveness of our recommendation system in providing users with valuable and personalized offerings.

Keywords: Large language models · Screenshots · User preferences · Recommendations

1 Introduction

In the evolving landscape of e-commerce, the need to understand user preferences for offering personalized content is increasingly crucial for online platforms to maximize revenue, build customer loyalty, and maintain competitive advantage. Typically, the nuanced information that can explain the user browsing behavior on these platforms is stored as weblogs [7]. Numerous state-of-the-art recommendation systems, employing strategies such as collaborative filtering,

Z. Wang and C. W. Tan (Eds.): PAKDD 2024 Workshops, LNAI 14658, pp. 32–43, 2024.
https://doi.org/10.1007/978-981-97-2650-9_3

content-based filtering, and hybrid methods, leverage the data mined through these weblogs to provide personalized recommendations. This data-driven approach ensures that recommendations align closely with user preferences, contributing to a more seamless and satisfying shopping journey and ultimately increasing overall platform revenue.

Raw weblogs contain a wealth of information on key browsing session details, including start and end times, visited pages, and clickstream data. Beyond capturing fundamental details, these weblogs delve into user-specific data, encompassing identifiers and IP addresses. However, interpreting the raw web log data directly can be challenging for non-experts, requiring them to navigate its complexity and dissect it to build recommendation models. Often, sophisticated data engineering techniques are necessary to extract the relevant features and datasets needed for training these models. With recent advancements in artificial intelligence, particularly within the domain of LLMs [11], a compelling case emerges for developing a principled framework that could present a superior solution to the current dependency on weblogs.

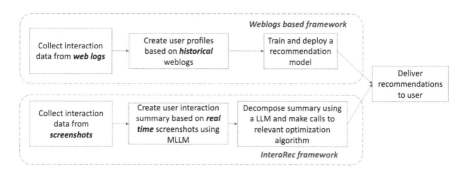

Fig. 1. A comparison of weblog-based framework and *InteraRec* framework (refer to Fig. 2 for full system description).

Inspired by the above, we propose an innovative approach leveraging information from screenshots of users' internet browsing activities. Our approach leverages MLLMs adept at seamless processing and generating content across diverse modalities. By opting for screenshots as inputs instead of relying on weblogs, the system benefits from heightened interpretability. The visual nature of screenshots provides a lucid and transparent representation of user actions, significantly enhancing the comprehension of the inferences made by the LLM. Unlike weblogs, which may introduce complexities associated with various features, the use of screenshots simplifies the inputs, offering a more straightforward and intuitive approach as shown in Fig. 1. Furthermore, this method facilitates concurrent, real-time recommendations, as screenshots can be captured almost instantaneously.

In this paper, we introduce *InteraRec*, a novel screenshot-based user recommendation system. We capture high-frequency screenshots of web pages as users navigate through them in real-time. Leveraging the capabilities of MLLMs, *InteraRec* processes these screenshots to extract meaningful insights into user behavior and execute pertinent optimization tools to deliver personalized recommendations. By seamlessly integrating visual data, language models, and optimization tools, *InteraRec* transcends the limitations of existing systems, promising a more personalized and effective recommendation system for users.

2   Related Work

In this study, we expand upon two key streams of research: (a) LLMs for user behavior modeling and (b) Tools and their integration with LLMs. We briefly discuss some of the related works below.

LLMs for User Behavior Modeling. LLMs are known for their capability to dynamically generate diverse facets of user profiles by analyzing their historical viewing and transaction data. Chen et al. [1] generated user profiles from TV viewing history to retrieve candidates from an item pool, subsequently leveraging LLM for item recommendations. Liu et al. [4] employed LLMs to generate user profiles, exploring topics and regions of interest based on user browsing history and integrated the inferred user profile to enhance recommendations. Unlike the aforementioned approaches that rely on text data from user history for recommendation generation, our novel proposal creates dynamic user profiles by leveraging screenshots of user interactions using MLLMs.

Tools and their Integration with LLMs: Researchers have made significant strides in using LLMs to tackle complex tasks by extending their capabilities to include planning and API selection for tool utilization. For instance, Schick et al. [8] introduced the pioneering work of incorporating external API tags into text sequences, enabling LLMs to access external tools. Shen et al. [9] proposed HuggingGPT, a novel framework leveraging LLMs as controllers to manage existing domain models for intricate tasks effectively. Lastly, Qin et al. [6] proposed a tool-augmented LLM framework that dynamically adjusts execution plans, empowering LLMs to complete each subtask using appropriate tools proficiently. Li et al. [3] introduced the Optiguide framework, leveraging LLMs to elucidate supply chain optimization solutions and address what-if scenarios. In contrast to the aforementioned approaches, *InteraRec* harnesses the power of LLMs to generate item recommendations by converting user behavior summaries into mathematical constraints that can be readily interpreted by an optimization solver.

3   The *InteraRec* Framework

Addressing the challenge of offering real-time recommendations to users is a multi-step process. It typically commences with thorough data collection and analysis, followed by the careful selection and training of appropriate user behavior models. After a rigorous evaluation, the trained model is deployed to generate optimal recommendations. The user is then empowered to make a decision, choosing whether to act upon or disregard the presented recommendations.

Our framework, *InteraRec*, depicted in Fig. 2, begins by systematically capturing screenshots of a user's browsing activity on a designated webpage at regular intervals. Subsequently, an MLLM, with its inferential capabilities, translates user interaction behavior into summary information comprising predefined keywords. This information is then processed by an LLM equipped with function-calling ability, transforming the summary into specific constraints and executing optimization tools following a validation check. The generated solutions are communicated to the user via the LLM through a user interface. The process outlined above is organized into multiple stages: 1) Screenshot generation, 2) Behavioral summarization, and 3) Response generation.

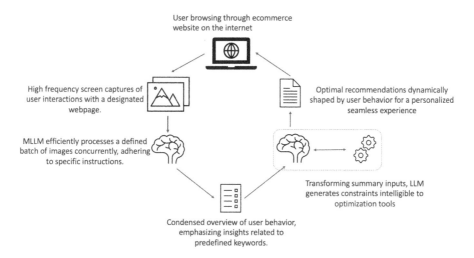

Fig. 2. Overview of *InteraRec* framework delivering real-time personalized recommendations

3.1 Screenshot Generation

InteraRec initiates an automated script to systematically capture high-frequency, high-quality screenshots of the web pages users navigate during a browsing session. Notably, the screenshots are confined to the user's current viewing area on the screen rather than encompassing the full scrollable content of each webpage. This allows for efficient targeting of the visible content that users actively engage with. *InteraRec* then stores these representative screenshots in a database for further processing and analysis.

3.2   Behavioral Summarization

In this stage, *InteraRec* sequentially processes a finite number of screenshots in real-time using an MLLM to analyze and provide detailed responses. Specifically, *InteraRec* instructs the MLLM to succinctly summarize user interaction behavior across predefined categories such as product characteristics, lowest and highest prices, brand preferences, product specifications, user reviews, comparisons, and promotions as shown in Fig. 3. This precise instruction is crucial to filter the information and capture elements directly aligned with user interests. Generic instructions like "describe the user behavior on screenshots" or "explain the differences in screenshots" yielded responses that lacked specificity and failed to reflect user preferences. If the model cannot deduce information for a given category, *InteraRec* generates a response indicating 'does not know' or 'not applicable.' Additionally, *InteraRec* ensures that the model presents information in a summarized JSON format, facilitating its use as a constraint or filter for subsequent processing.

```
response = client.chat.completions.create(
model="gpt-4-vision-preview"
messages=[
{
"role": "user"
"content": [
{
"type": "text"
"text": "What can you infer from the images below \
with regards to a user preference in the following categories? \
Product Characteristics, \
Lowest Price, \
Highest Price, \
Brand Preference, \
Product Specifications, \
User Reviews and Testimonials, \
Comparisons, \
Promotions. \
Write a response that contains the above information. \
if any of the categorical information is unavailable, \
mark it as not available "}...],...}
```

Fig. 3. Guidelines for MLLM to generate a summary of user interactions using predefined keywords

3.3   Response Generation

The keyword-based summary generated from the previous stage is rich in information, forming a valuable resource for generating recommendations. The relevant information to be extracted from the summary depends on the capabilities of recommendation methods that can incorporate them. In this study, *InteraRec* harnesses the *InteraSSort* framework [2] to deconstruct the user behavior summary into relevant constraints and uses them to solve an assortment optimization problem to generate optimal recommendations. Specifically, *InteraRec* leverages the function-calling capabilities of the LLM to decompose the user interaction summary into relevant constraints using appropriate functions as shown in Fig. 4. Subsequently, rigorous validation checks are conducted, encompassing range and consistency assessments of the decomposed constraints. *InteraRec* maintains an extensive database containing parameters for discrete choice models estimated using historical purchase transactions. By leveraging these choice model parameters and additional constraints as arguments, *InteraRec* executes optimization scripts employing tools like optimization solvers to generate optimal solutions. Ultimately, *InteraRec* empowers the LLM to incorporate these results as input, generating user-friendly personalized product recommendations.

```
function_descriptions = [
    {
    "name": "get_user_recommendations",
    "description": "Generate dynamic  recommendations based \
                    the summary of user behavior",
    "parameters": {
    "type": "object", "properties": {
        "lowest_price": {
            "type": "integer",
            "description": "get lowest price preference of user."
        ,},
        "highest_price": {
            "type": "integer",
            "description": "get highest price preference of user."
        ,},
        "color": {
            "type": "string",
            "description": "get color preference of user",
        }, }, }, }, ]
```

Fig. 4. Prospective function for decomposing user interaction summaries.

4   Illustration

In this section, we discuss the components needed to run our experiments, followed by some illustrative examples.

4.1   Assortment Planning

The assortment planning problem involves choosing an assortment among a set of feasible assortments ($\mathcal{S}$) that maximizes the expected revenue. Consider a set of products indexed from 1 to n with their respective prices being $p_1, p_2, \cdots p_n$. The revenue of the assortment is given by $R(S) = \sum_{k \in S} p_k \times \mathbb{P}(k|S)$ where $S \subseteq \{1, ..., n\}$. The expected revenue maximization problem is simply: $\max_{S \in \mathcal{S}} R(S)$. Here $\mathbb{P}(k|S)$ represents the probability that a user chooses product k from an assortment S and is determined by a choice model.

The complex nature of the assortment planning problem requires the development of robust optimization methodologies that can work well with different types of constraints and produce viable solutions within reasonable time frames. In our experiments, we adopt a series of scalable efficient algorithms [10] for solving the assortment optimization problem.

4.2   Multinomial Logit (MNL)

The MNL model [5] is one of the most extensively studied discrete choice models and is frequently utilized across various marketing applications. The parameters of the MNL model are represented by a vector $\mathbf{v} = (v_0, v_1, \cdots v_n)$ with $0 \leq v_i \leq 1$ $\forall i$. Parameter v_i, $1 \leq i \leq n$, captures the user preference for purchasing product i. Under this model, the probability that a user chooses product k from an assortment S is given by $\mathbb{P}(k|S) = v_l/(v_0 + \sum_{k' \in S} v_{k'})$.

4.3   LLM

We employ the `gpt-4-vision-preview` (GPT-4V) and `gpt-3.5-turbo` variants from the GPT model series as our MLLM and LLM, respectively. Both the models are publicly accessible through the OpenAI API[1].

4.4   Illustrative Examples

Example 1: Through this illustrative example, we present the dynamic process of generating personalized, real-time recommendations for users navigating the *Asos* fashion apparel website. The recommender system (*InteraRec*) initiates by capturing screenshots every 1 s, as depicted in Fig. 5, and stores them in the database. Following this, *InteraRec* employs GPT-V to process these images in batches of 10 due to its input token size limitation and generate a summary

---

[1] https://platform.openai.com/.

Fig. 5. Six screen captures showcasing user interactions with the Asos.com website, each captured at a 1-s interval.

Given the images shown, it's important to note that while certain inferences can be made about product preferences and price sensitivity among other things, these are assumptions based on a set of images featuring green dresses of different styles and prices. Here is a possible interpretation in JSON format: "json { *"Product Characteristics"*: "User appears to be interested in dresses, specifically in shades of green, with various styles including mini, midi, and maxi lengths.", *"Lowest Price"*: "18.00$ for the JDY puff sleeve mini smock dress in bright green", "Highest Price": "144.00$ for the Y.A.S Bridesmaid satin cami maxi dress with ruching detail in deep teal green", *"Brand Preference"*: "Variety of brands shown, such as Y.A.S, Wednesday's Girl, Nobody's Child, suggesting no clear preference for a single brand.", *"Product Specifications"*: "Not Available", *"User Reviews and Testimonials"*: "Not Available", *"Comparisons"*: "The user may be comparing dresses based on style and discounts, with several items showing reduced prices.", *"Promotions"*: "Several dresses are listed with percentage discounts, indicating that promotions may influence user preferences." } "

The response indicates an interest in green dresses, sensitivity to promotions (as indicated by discounted items), a range of price points indicating a potentially broader budget range or comparison shopping, and a variety of brands.

Fig. 6. Extracted summary from *Asos* website activities revealing a user preference for green dresses, and price sensitivity. (Color figure online)

based on provided instructions, as illustrated in Fig. 6 (see Appendix A for additional example). In our approach, we focus on extracting the following set of constraints, specifically the color and price range of products. Subsequently, the system parses the input, extracting details such as color ('green'), lowest price as '18$'and highest price as '144$') by leveraging the function-calling ability of the `gpt-3.5-turbo` model. Using the parsed information as arguments, *InteraRec* executes the MNL optimization script and communicates outcomes through the interface. We can observe the impact of incorporating user behavior in the form of updated recommendations limited to green apparel, as shown in the Fig. 7.

This methodical, step-by-step approach ensures a consistent refinement of recommendations through continuous analysis of successive image batches.

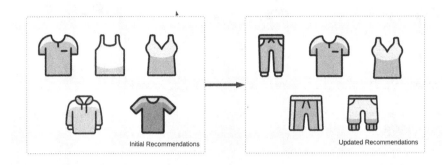

Fig. 7. *InteraRec* Real-time recommendations incorporating the dynamic user behavior.

Fig. 8. Nine screen captures showcasing user interactions with the Nike.com website, each captured at a 4-s interval.

Example 2: In this example, we present the dynamic process of generating personalized, real-time recommendations for users navigating the *Nike* website. The recommender system (*InteraRec*) initiates by capturing screenshots every 4 s, as depicted in Fig. 8, and stores them in the database. Following this, *InteraRec* processes these images in batches of 10 to generate a summary based on provided instructions as illustrated in Fig. 9. Subsequently, *InteraRec* extracts the following set of constraints: brand preference and price range of products. The system then parses the input, extracting details such as brand preference 'Air Jordan', lowest price as '63.97$' and highest price as '180$'). Using the parsed information as arguments, *InteraRec* executes the MNL optimization script and communicates outcomes through the interface.

"json *"Product Characteristics"*: "The products featured are mostly casual and sports footwear, with a focus on style and functionality.", *"Lowest Price"*: "$63.97, as seen on the Nike Waffle Debut Womens Shoes.", *"Highest Price"*: "$180, as seen on the Air Jordan 1 High G Mens Golf Shoes.", *"Brand Preference"*: "The brand showcased is Nike, including its Air Jordan line, indicating a preference for this brand.", *"Product Specifications"*: "Gender": "Men, Women, and Unisex options available", "Color": "Multiple color options are available for selection", "Best For": "options like Cold Weather, Dry Weather Conditions, Wet Weather Conditions are available, although specific preferences are not indicated in the images." , *"User Reviews and Testimonials"*: "Not available", *"Comparisons"*: "The images showcase different models and prices, suggesting that comparisons can be made within the selection, but specific comparative analysis is not displayed.", *"Promotions"*: "Promotions are visible, such as WINTER SALE: UP TO 50% OFFand discounts on specific products.""

Fig. 9. Extracted summary from *Nike* website activities revealing a user preference for a specific shoe brand and price sensitivity.

5 Conclusion

In this paper, we introduce *InteraRec*, an interactive framework designed to craft personalized recommendations for users browsing an e-commerce platform. Overcoming the limitations of extensive weblog engineering, *InteraRec* utilizes screenshots to capture and understand user behavior. Harnessing the capabilities of both LLM and MLLMs, *InteraRec* leverages insights from the screenshots and intelligently executes suitable optimization tools, translating user behavior into concise, easily interpretable recommendations for users. Overall, *InteraRec* heralds a new era in recommendation systems driven by visual data.

A   Appendix

(See Figs. 10 and 11).

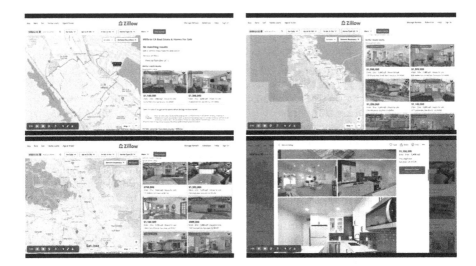

Fig. 10. Four screen captures showcasing user interactions with the Zillow.com website, each captured at a 10-second interval.

"Based on the provided images, it appears that the user is searching for real estate properties for sale on a website, likely with certain preferences. The images show a website with real estate listings that include price ranges, the number of bedrooms and bathrooms, square footage, and location. Using the data visible in the images, I will attempt to create JSON-formatted data that represents the users inferred preferences within the limitations of the images. "json *"Product Characteristics"*: "Location": "Millbrae, CA; South San Francisco, CA; Pacifica, CA; San Jose, CA", "Property Type": "House", "Bedrooms": "3+", "Bathrooms": "2+" , *"Lowest Price"*: "$799,888", *"Highest Price"*: "$1,350,000", *"Brand Preference"*: "not available", *"Product Specifications"*: "Square Footage": "Low": ",054 sqft", "High": "3,040 sqft" , *"User Reviews and Testimonials"*: "not available", *"Comparisons"*: "Available": true, "Criteria": "Price, Location, Size, Bedrooms, Bathrooms" , *"Promotions"*: "Price Cut": "$20,000 (Oct 31) on a property listed for $989,000" ''Please note that the information above is constructed from the context provided by the images and might lack accuracy in details that are not visible or provide insufficient context. Some categories like Brand Preferenceánd Úser Reviews and Testimonialsdo not apply in the context of real estate listings and therefore are marked as "not available." Also, the promotion information is based on limited data indicating a price cut on a specific property, as seen in the images."

Fig. 11. Extracted summary of user preferences shaped by activities on the Zillow.com website.

InteraRec: Interactive Recommendations Using MLLMs   43

References

1. Chen, Z.: PALR: personalization aware LLMs for recommendation. arXiv preprint arXiv:2305.07622 (2023)
2. Karra, S.R., Tulabandhula, T.: InteraSSort: interactive assortment planning using large language models. arXiv preprint arXiv:2311.12241 (2023)
3. Li, B., Mellou, K., Zhang, B., Pathuri, J., Menache, I.: Large language models for supply chain optimization. arXiv preprint arXiv:2307.03875 (2023)
4. Liu, Q., Chen, N., Sakai, T., Wu, X.M.: A first look at LLM-powered generative news recommendation. arXiv preprint arXiv:2305.06566 (2023)
5. Luce, R.D.: Individual Choice Behavior: A Theoretical Analysis. Courier Corporation (2012)
6. Qin, Y., et al.: Tool learning with foundation models. arXiv preprint arXiv:2304.08354 (2023)
7. Rosenstein, M.: What is actually taking place on web sites: e-commerce lessons from web server logs. In: Proceedings of the 2nd ACM Conference on Electronic Commerce, pp. 38–43 (2000)
8. Schick, T., et al.: Toolformer: language models can teach themselves to use tools. arXiv preprint arXiv:2302.04761 (2023)
9. Shen, Y., Song, K., Tan, X., Li, D., Lu, W., Zhuang, Y.: HuggingGPT: solving AI tasks with ChatGPT and its friends in HuggingFace. arXiv preprint arXiv:2303.17580 (2023)
10. Tulabandhula, T., Sinha, D., Karra, S.: Optimizing revenue while showing relevant assortments at scale. Eur. J. Oper. Res. **300**(2), 561–570 (2022)
11. Zhao, W.X., et al.: A survey of large language models. arXiv preprint arXiv:2303.18223 (2023)

Research on Dynamic Community Detection Method Based on Multi-dimensional Feature Information of Community Network

Kui Hu[1](✉), Zhenyu Zhang[1,2], and Xiaoming Li[3]

[1] School of Computer Science and Technology, Xinjiang University, Urumqi 830017, China
2630008231@qq.com

[2] Xinjiang Key Laboratory of Multilingual Information Technology, Urumqi 830017, China
zhangzhenyu@xju.edu.cn

[3] College of International Business, Zhejiang Yuexiu University, Shaoxing, China

Abstract. With the continuous development of technology, we have the ability to fully record all aspects of data information of every individual in the society, so how to utilize this information to create greater value is becoming more and more important. Compared with the traditional static community detection, the study of dynamic community detection is more in line with the real situation in the society. Thus, in this paper, a method that can utilize the information of diversified dynamic community networks is proposed, i.e., Dynamic Community Detection Method based on Multidimensional Feature Information of Community (Dcdmf), which utilizes neural networks with strong learning and adaptive capabilities, the ability to automatically extract useful features and process complex data, and the ability to process the graph nodes and the data between the nodes of the dynamic community network, and the ability to real-time adjust the current community representation data based on historical information, and record the current community representation data for the next moment of community data. The experimental results in the paper show that the method has a certain degree of effectiveness.

Keywords: dynamic community detection · neural networks · multidimensional features · historical information

1 Introduction

The structure of community network is a graph structure that abstracts the relationship between people in the community into nodes and edges, and this expression method facilitates the value and intuitive discovery of one small community structure in a complex community network, which originates from Graph theory [1]. The community structure refers to the nodes in the network can be

Z. Wang and C. W. Tan (Eds.): PAKDD 2024 Workshops, LNAI 14658, pp. 44–56, 2024.
https://doi.org/10.1007/978-981-97-2650-9_4

divided into multiple subgroups, the node edges within the group are relatively tight, and the node connecting edges between the groups are relatively sparse [2].

Research on the problem of discovering the structure of communities helps to understand the function of community networks and their evolution and organization information. Early research on community networks focused on static community structure, but in the real world, the structure of the community network is constantly changing over time, so dynamic community detection and evolution prediction research has become a current research hotspot. About the application of dynamic community research: in citation network, it helps to understand the current research hotspots and possible new research directions in the future; in social entertainment network, it helps to understand the dissemination of information and help ordinary people to find the people they may know; in e-commerce, it can be more accurate advertising to help merchants sell their goods better, etc. [3].

2   Related Work

Dynamic community detection needs to take into account that the community structure changes over time [4]. Dakiche N et al. [5] categorized the dynamic community detection algorithms into two main classes, the first class is the algorithms that deal with snapshots of the community, which are less real-time and have lower arithmetic requirements, and the second class is the ones that deal with the events of the community [6], which are more real-time and have higher arithmetic requirements.

The first class of algorithms dealing with snapshots of association networks is due to the early research on dynamic association detection, most of which transform static association detection algorithms by introducing temporality, such as, Newman's fast algorithm [7], Kernighan-Lin [8], Synwalk [9], Infomap [10], and Walktrap [11], etc. The DyCPM [12] is based on the CPM [13] introduced temporal ordering transformation, CPM for each community network snapshot, which provides traversing the complete subgraphs in the network to find all the possible seed communities, and constantly merge multiple communities into a larger community until it is no longer possible to merge the community communities or reach the pre-determined maximum size of the community. Meanwhile, with the continuous development of deep learning [14], some people have applied graph neural network related models to community detection, which can effectively capture the complex relationship and topology between nodes within a community snapshot, thus improving the accuracy and expressiveness of the community structure, and by introducing temporal ordering to this kind of model, it can also be applied to dynamic community detection, such as, LGNN [15], GAT [16] and GraphSAGE [17]. Evolutionary clustering [18] based community detection algorithms, which explore the evolution of community structure in dynamic networks through an iterative process of evolutionary algorithms, discover the optimal community partitioning of the network at different time points by optimizing the fitness function to reveal the temporal evolutionary pattern of the community, such as, FacetNet [19], DYNMOGA [20], MODPSO [21] and DLPAE [22].

The second class of approaches to deal with community events, incremental clustering-based community detection algorithms, which effectively detect the community structure by utilizing the incremental information of the network structure over time, thus reducing the computational complexity and maintaining the accuracy of the community detection, such as, CUT [23], DABP [24], QCA [25], and IA-MCS [26]. CUT reduces the computational complexity and maintains the accuracy of the community detection by dynamically updating the representative tree structure in real time to adapt to changes in the data, thus enabling it to efficiently handle clustering tasks in data streams or dynamic environments.

3   Problem Description

Currently, the dynamic community detection algorithms introduced by static community detection and transformed from temporal sequencing roughly include clustering-based algorithms, modularity-based algorithms, and complete subgraph-based algorithms, etc. These algorithms only consider the information of the graph's topology or degree centrality, and do not make use of node's feature information; and most of the methods use the graph's neighbour matrix, which is not conducive to dealing with large-scale graphs, such as For example, if an undirected graph has 1 million nodes and uses one byte to record the information of an edge, then the memory consumption is about 931.3G (the actual situation may be larger), if the adjacency matrix can be changed to an adjacency table, the memory consumption will be greatly reduced. Dynamic community detection needs to consider community evolution events [5,12,27–29], but many current methods do not consider historical information; although incremental clustering-based methods are a method that can make use of historical information, they do not make use of node feature information, and they are prone to fall into local optimums, and their accuracy decreases when there are large changes in the community structure.

In summary, the problems of the current dynamic community detection methods can be attributed to the fact that they cannot effectively use the feature information of the graph nodes or edges as well as the historical community structure information, which leads to the lack of accuracy of the community detection structure.

4   Algorithmic Model

In this subsection, a dynamic community detection method based on the multidimensional feature information of communities (Dcdmf) is presented. The data processed by this model is a graph in which all the node features are processed into a matrix, where each row represents the multidimensional feature information of a node; the edge information is processed into an adjacency table. The moment t is used in the text to represent the index of a snapshot. The algorithm model diagram, as in Fig. 1.

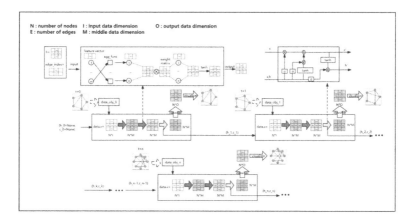

Fig. 1. Model diagram of dcdmf

4.1 Feature Processing

In the Dcdmf model, Graph Convolutional Network (GCN) [30,31] is used to extract the feature information among nodes through the convolution operation and the structural information of the graph. In the method proposed in this paper, a 2-layer graph convolutional network is used to process the node features, and the choice of using two layers here is to prevent too many layers of the network from causing overfitting of the nodes and insufficient convolution of the node information resulting in inaccurate results due to too few layers of the network. Each layer of the network has the following equation when processing the input at moment t.

$$H^{l+1} = \sigma(\overline{D}^{-\frac{1}{2}} \overline{A} \overline{D}^{\frac{1}{2}} H^l W^l) \tag{1}$$

In Eq. 1, is derived from the adjacency matrix of the input data plus a unit diagonal matrix as Eq. 2.

$$\overline{A} = A + I \tag{2}$$

There are two cases of H in Eq. (1), the first is the node characteristics of the input graph data, and the second is the convolution result of the output of the first layer of the network and the matrix type data obtained by the operation of the activation function σ. The activation function here is tanh, which is a nonlinear function that compresses the input values to the range of $[-1, 1]$ by adjusting the weights of positive and negative exponents, and can be used in neural networks by its It is widely used to increase the nonlinear expression ability of the model and speed up the convergence of the model. Tanh is formulated as Eq. 3. $\overline{D}$ is the degree-diagonal moment of $\overline{A}$. The equation is as Eq. 4.

$$tanh(x) = \frac{e^x - e^{-1}}{e^x + e^{-1}} \tag{3}$$

$$\overline{D} = Diag(\sum_j \overline{A}[i][j]) \tag{4}$$

In Eq. (1), W represents the weight matrix, which in graph convolutional networks serves to linearly transform the feature matrices of the nodes to capture the feature information among the nodes. The feature information of the first and second layer of nodes centred on that node stratified by breadth priority is obtained by feature processing each node in the graph, which facilitates processing in subsequent steps.

4.2   Reference to Historical Information

For the result matrix after feature processing, it is also necessary to regulate the state of the matrix with reference to the information of the community structure before the moment t in order to optimise the result. In this paper, the proposed method uses the Long Short-Term Memory (LSTM) neural network [32] to regulate the feature-processed matrix, which was proposed by S. Hochreiter and J. Schmidhuber in 1997, and Alex Graves proposed a Gated Recurrent Unit (GRU) [33] on top of LSTM, although its structure is simpler and has fewer parameters. Its structure is more concise and the number of parameters is less, its performance may be degraded compared to LSTM when dealing with particularly long sequences. The core structure of LSTM consists of three gating units: forgetting gate, input gate and output gate. These gating units are composed of multiple neurons, which process the data through weighting calculation and activation function.

$$Input\ gate : I_t = \overline{\sigma}(W_{xi} * x_t + W_{hi} * h_{t-1} + b_i) \tag{5}$$

In Eq. (5), I_t denotes the input gate at moment t, W_{xi} and W_{hi} denote the weight matrix of the input gate and the hidden state of the previous step, x_t denotes the input vector at moment t, h_{t-1} denotes the hidden state at moment t − 1, b_i denotes the input gate bias, and $\overline{\sigma}$ denotes the sigmoid activation function, which maps the input values between 0 and 1. There is the Eq. 6.

$$sigmoid(x) = \frac{1}{1 + e^{-x}} \tag{6}$$

$$Forgetting\ gate : f_t = \overline{\sigma}(W_{xf} * x_t + W_{hf} * h_{t-1} + b_f) \tag{7}$$

In Eq. (7), f_t denotes the forgetting gate at moment t, W_{xf} and W_{hf} denote the weight matrices of the forgetting gate and the hidden state of the previous step, and b_f denotes the forgetting gate bias.

$$Memory\ cell\ status : c_t = f_t * c_{t-1} + i_t * \sigma(W_{xc} * x_t + W_{hc} * h_{t-1} + b_c) \tag{8}$$

In Eq. (8), c_t and c_{t-1} denote the memory cell states at moments t and t − 1, respectively, σ is Eq. (3), W_{xc} and W_{hc} denote the input gates of the memory cell and the weight matrix of the hidden state in the previous step, and b_c denotes the bias of the memory cell.

$$Output\ gate : o_t = \overline{\sigma}(W_{xo} * x_t + W_{ho} * h_{t-1} + b_o) \tag{9}$$

$$Hidden\ status : h_t = o_t * \sigma(c_t) \tag{10}$$

In Eq. (9), o_t denotes the output gate at moment t, W_{xo} and W_{ho} denote the weight matrix of the output gate and the hidden state of the previous step, and b_o denotes the bias of the output gate.

Overall, the Dcdmf model needs to process the node feature information of the pending community snapshot into a feature matrix and the topology structure into an adjacency table before each run. Subsequently, through feature processing, each node obtains representation information of all nodes within a path length range of 2 starting from it. Next, by referencing historical club information, adjust the current club information and scale the length of the node vector to the number of clubs. Finally, based on which dimension the node vector is most likely to belong to, it is divided into corresponding sub communities.

5   Experiments

This section describes the experiment-related datasets, evaluation metrics, and results; the hardware conditions used for the experiments in this paper are: the graphics card is a Tesla V100-SXM2-16 GB, the processor is an Intel(R) Xeon(R) Gold 6148, and the RAM is 64 GB.

5.1   Datasets

Synthetic Dataset. The generation method consists of three steps: the first step is to create a link matrix A with 5 groups (which has 5000 nodes and 20000 edges), and connect one set of edges to another set of edges. Step 2, create a content matrix C with 5 words, and one word can appear in multiple groups. Step three, select the probability of rewiring p = 0.75 and the probability of whether the node is reorganized q = 0.1, and generate 3 timestamps.

Dgraphfin Dataset. Dgraphfin [34] is a directed, unweighted dynamic graph consisting of millions of nodes and edges, which contains four types of users. It represents a social network between Finvolution Group users, where nodes represent Finvolution users, and edges from one user to another imply that a user considers another user as an emergency contact. Each user uses a 17-dimensional feature vector to describe its relevant feature information, which provides 821 snapshots (in days), 3,700,500 nodes, and 4,300,999 edges.

5.2   Evaluation Indicators

Modularity. Modularity [35] is used to evaluate the results of community division by calculating the difference between intra-community connection density and inter-community connection density, the higher the value of modularity, the better the effect of community division, and the range of its value is $[-1, 1]$.

$$Modularity = \frac{1}{2m} \sum_{i,j \in C_n} (A_{ij} - \frac{k_i k_j}{2m}) \qquad (11)$$

In Eq. (11), A_{ij} denotes the number of edges between node i and node j, m denotes the number of edges, $\frac{k_i k_j}{2m}$ denotes the expected value of the number of edges between node i and node j in case of randomly placed edges, and C_n denotes the nth community in the graph data.

Jaccard Similarity. It measures the similarity of communities by calculating the ratio of the size of the intersection of two communities to the size of their concatenation; the closer the JACCARD value is to 1, the more similar the two communities are.

$$Jaccard(A, B) = \frac{|A \cap B|}{|A \cup B|} \tag{12}$$

Consumption Time. The average time, in seconds, spent by the algorithm to process a single community network snapshot.

For the three evaluation indicators mentioned above, they are replaced by Q (modularity), J (Jaccard similarity), and T (time consumed) in the following, and EI in the table stands for evaluation indicator.

5.3   Comparison Algorithm

To verify the effectiveness of using multidimensional feature information, the following 5 algorithms were selected for comparison.

Clustering Algorithm. This paper chose KMeans [36] for experimentation, the algorithm in the form of k clusters to group the data, through iteration to optimise the objective function, so that each data point to the centre of mass of the cluster to which it belongs to the minimum of the sum of the squared distance.

Algorithm Based on Modularity. In this paper, DyLouvain is selected for experiments, which is adapted from Louvain [37] by introducing temporal ordering. Louvain is a community discovery algorithm based on modularity optimisation, which iteratively merges nodes in a network to reveal the community structure in the network.

Label Propagation Based Algorithm. In this paper, we have chosen to experiment with LPA [38], which is an algorithm that identifies the network community structure by constantly updating the labels of each node, where a node's label is determined based on the labels of its neighbouring nodes, and similar nodes end up with the same labels, thus forming a community.

Methods Based on Dynamic Network Community Evolution Models. DECS [39] is an association detection method based on dynamic network community evolution models. The method discovers the community structure by tracking the membership changes and interactions of each community.

Content and Link Based Approach. Chimera [40] which is a method for detecting and predicting community structure using links and content in a network is an efficient algorithm based on the decomposition of sharing matrices and the uniform nature of the embedding of this algorithm allows the use of clustering algorithms on the corresponding representations.

5.4 First Experiment

Three consecutive snapshots of the synthetic dataset are processed, the experiment is averaged over five times and the metrics are evaluated, and the experimental results are shown in Table 1.

Table 1. Experimental results on the synthetic dataset

ALG	EI	1	2	3	ALG	EI	1	2	3
KMeans	Q	0.0002	0.0003	0.0004	Decs	Q	0.4452	0.4464	0.4471
	J	0.1083	0.1098	0.1073		J	0	0.2	0.2
	T	0.30	0.29	0.27		T	37.05	36.39	40.67
DyLouvain	Q	0.4452	0.4464	0.4471	Chimera	Q	0.4425	0.4427	0.4427
	J	0.4	0.0	0.2		J	0.9791	0.9093	0.9621
	T	2.77	2.70	3.25		T	0.253	0.283	0.282
LPA	Q	0	0	0	Dcdmf	Q	0.4415	0.4428	0.4437
	J	0.04	0.04	0.04		J	0.9932	0.9932	0.9936
	T	0.31	0.36	0.32		T	0.20	0.15	0.18

In the synthetic dataset experiment, KMeans, DyLouvain, Decs, and Chimera detected 5 communities respectively, while LPA only presented 1 community. Dcdmf also found 5 communities in each experiment. This preliminary reveals the performance of various algorithms in community detection tasks.

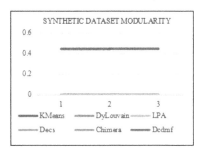

Fig. 2. Modularity of first experiment

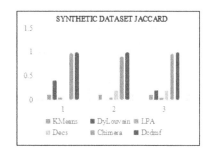

Fig. 3. Jaccard of first experiment

In the experimental results, we observed that the modularity values of Dcdmf, Chimera, Decs, and DyLouvain were similar and significantly better than KMeans and LPA, with the modularity values of the latter two approaching 0. This trend is presented in Fig. 2 in data visualization. In addition, in terms of JACCARD indicators, Dcdmf and Chimera performed better than other methods, while LPA performed the lowest. It is worth noting that the JACCARD index of Dcdmf is higher than that of Chimera, as shown in Fig. 3. In addition, we also observed that each algorithm has different time consumption for processing a single snapshot, with Dcdmf algorithm having the shortest time consumption, Chimera taking second place, and Decs having the longest time consumption. This information is clearly displayed in the data visualization in Fig. 4.

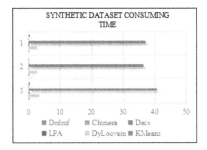

Fig. 4. Consuming time of first experiment

5.5   Second Experiment

Due to the excessive number of community snapshots in the DGRAPHFIN dataset (821 snapshots in total), the 821 snapshots were divided into three groups in the experiment (the number of snapshots was 274, 274, 273), and the results of the experiment are shown in Table 2.

Table 2. Experimental results on the dgraphfin dataset

ALG	EI	1	2	3	ALG	EI	1	2	3
KMeans	Q	−0.017	−0.058	−0.076	Decs	Q	0.003	0.0038	0.0034
	J	0.074	0.104	0.060		J	0.065	0.099	0.098
	T	44.30	42.40	36.80		T	784.47	1045.44	978.79
DyLouvain	Q	0.99	0.99	0.99	Chimera	Q	0.0408	0.0486	0.0342
	J	0	0	0		J	0.152	0.153	0.147
	T	417.40	603.40	457.10		T	6.20	7.31	7.02
LPA	Q	0.88	0.83	0.87	Dcdmf	Q	0.00038	0.00028	0.00068
	J	0	0	0		J	0.265	0.268	0.283
	T	51.80	64.40	46.90		T	4.11	4.46	4.48

In the experiment on the DGraphFin dataset, KMeans detected 4, 4, and 4 communities respectively; DyLouvain discovered 473220, 436012, and 399698 clubs respectively; And LPA discovered 617534, 705790, and 549963 communities respectively. In contrast, Decs and Chimera detected 4 communities in each experiment, while Dcdmf also found 4 communities in each experiment.

Analysis shows that DyLouvain and LPA have the highest modularity, but the number of discovered communities far exceeds the set. The visualization of experimental data confirms this phenomenon. To verify the credibility of the Dcdmf results, we directly calculated the modularity of three sets of snapshots on the Dgraphfin dataset, as shown in Table 3. The results are similar to those calculated by Dcdmf, confirming its reliability.

Table 3. Real modularity of dgraphfin dataset

EI	1	2	3
Q	0.00040	0.00023	−0.00041

In terms of JACCARD similarity, the Dcdmf algorithm performs the best, with a 10% improvement compared to the second ranked Chimera algorithm. It is worth noting that the calculation results of DyLouvain algorithm and LPA algorithm are 0, because the number of communities discovered by these two algorithms has far exceeded the four types of communities calibrated by the Dgraphfin dataset. In addition, Dcdmf also performs the best in terms of the average time consumed by the algorithm to process a single snapshot.

6   Conclusion

In this study, we propose a dynamic community detection method based on multidimensional feature information of communities (Dcdmf). Related experiments have shown that this method can effectively handle the multidimensional feature information of graph nodes, fully utilize the information of historical snapshots, and demonstrate good real-time performance. This study emphasizes that research on dynamic community detection should not only focus on graph topology, node centrality, and node features. Future research directions should improve the accuracy of detection results through diversified information. Of course, dynamic community detection still faces some challenges at present, such as the processing of community networks containing a large number of graph nodes, and the fact that most of the dynamic community network data on the network currently does not contain information such as labels.

Acknowledgments. This work was supported in part by the National Science Foundation of China (62272311), National Key R & D Program of China (2018YFC0831005), Science and Technology Support project of Tianjin Eco-City of China (STCKJ2020-WRJ), and Finance science and technology project of the 12th Division of Xinjiang Construction Corps of China (SR202103).

References

1. Bollobás, B.: Modern Graph Theory. Springer, New York (1998). https://doi.org/10.1007/978-1-4612-0619-4
2. Girvan, M., Newman, M.E.J.: Community structure in social and biological networks. Proc. Natl. Acad. Sci. **99**(12), 7821–7826 (2002)
3. Gasparetti, F., Micarelli, A., Sansonetti, G.: Community detection and recommender systems. In: Alhajj, R., Rokne, J. (eds.) Encyclopedia of Social Network Analysis and Mining, pp. 330–343. Springer, New York (2018). https://doi.org/10.1007/978-1-4939-7131-2_110160
4. Qiu, B., Ivanova, K., Yen, J., Liu, P.: Behavior evolution and event-driven growth dynamics in social networks. In: 2010 IEEE Second International Conference on Social Computing, Minneapolis, pp. 217–224. IEEE (2010). https://doi.org/10.1109/SocialCom.2010.38
5. Dakiche, N., Tayeb, F.B., Slimani, Y., Benatchba, K.: Tracking community evolution in social networks: a survey. Inf. Process. Manag. **56**, 1084–1102 (2019)
6. Greene, D., Doyle, D., Cunningham, P.: Tracking the evolution of communities in dynamic social networks. In: 2010 International Conference on Advances in Social Networks Analysis and Mining, Odense, pp. 176–183. IEEE (2010). https://doi.org/10.1109/ASONAM.2010.17
7. Newman, M.E.J.: Fast algorithm for detecting community structure in networks. Phys. Rev. E **69**(6), 066133 (2004)
8. Kernighan, B.W., Lin, S.: An efficient heuristic procedure for partitioning graphs. The Bell Syst. Tech. J. **49**(2), 291–307 (1970)
9. Toth, C., Helic, D., Geiger, B.C.: Synwalk: community detection via random walk modelling. Data Min. Knowl. Disc. **36**, 739–780 (2022)
10. Rosvall, M., Bergstrom, C.T.: Maps of random walks on complex networks reveal community structure. Proc. Natl. Acad. Sci. **105**, 1118–1123 (2007)
11. Pons, P., Latapy, M.: Computing communities in large networks using random walks. In: Yolum, P., Gungör, T., Gurgen, F., Özturan, C. (eds.) ISCIS 2005. LNCS, vol. 3733, pp. 284–293. Springer, Heidelberg (2005). https://doi.org/10.1007/11569596_31
12. Palla, G., Barabási, A.L., Vicsek, T.: Quantifying social group evolution. Nature **446**, 664–667 (2007)
13. Palla, G., Derényi, I., Farkas, I., Vicsek, T.: Uncovering the overlapping community structure of complex networks in nature and society. Nature **435**, 814–818 (2005)
14. Goyal, P., Ferrara, E.: Graph embedding techniques, applications, and performance: a survey. Knowl.-Based Syst. **151**, 78–94 (2018)
15. Chen, Z., Li, L., Bruna, J.: Supervised community detection with line graph neural networks Machine Learning. ArXiv (2017)
16. Velickovic, P., Cucurull, G., Casanova, A., Romero, A., Lió, P., Bengio, Y.: Graph attention networks. In: International Conference on Learning Representations (2018)
17. Hamilton, W.L., Ying, Z., Leskovec, J.: Inductive representation learning on large graphs. In: Proceedings of the 31st International Conference on Neural Information Processing Systems, pp. 1025–1035. Curran Associates Inc., Red Hook (2017)
18. Chakrabarti, D., Kumar, R., Tomkins, A.: Evolutionary clustering. In: Proceedings of the 12th ACM SIGKDD International Conference on Knowledge Discovery and Data Mining, pp. 554–560. Association for Computing Machinery, New York (2006)

19. Lin, Y., Chi, Y., Zhu, S., Sundaram, H., Tseng, B.L.: FacetNet: a framework for analyzing communities and their evolutions in dynamic networks. In: Proceedings of the 17th International Conference on World Wide Web, pp. 685–694. Association for Computing Machinery, New York (2008)
20. Folino, F., Pizzuti, C.: An evolutionary multiobjective approach for community discovery in dynamic networks. IEEE Trans. Knowl. Data Eng. **26**(8), 1838–1852 (2014)
21. Li, H., Yin, Y., Li, Y., Zhao, Y., Wang, G.: Large-scale dynamic network community detection by multi-objective evolutionary clustering. J. Comput. Res. Dev. **56**(2), 281–292 (2019)
22. Liu, K., Huang, J., Sun, H., Wan, M., Qi, Y., Li, H.: Label propagation based evolutionary clustering for detecting overlapping and non-overlapping communities in dynamic networks. Knowl.-Based Syst. **89**, 487–496 (2015)
23. Ma, H., Huang, J.: CUT: community update and tracking in dynamic social networks. In: Social Network Mining and Analysis (2013)
24. Li, X., Wu, B., Guo, Q., Zeng, X., Shi, C.: Dynamic community detection algorithm based on incremental identification. In: 2015 IEEE International Conference on Data Mining Workshop, Atlantic City, pp. 900–907. IEEE (2015). https://doi.org/10.1109/ICDMW.2015.158
25. Nguyen, N.P., Dinh, T.N., Xuan, Y., Thai, M.T.: Adaptive algorithms for detecting community structure in dynamic social networks. In: 2011 Proceedings IEEE INFOCOM, Shanghai, pp. 2282–2290. IEEE (2011). https://doi.org/10.1109/INFCOM.2011.5935045
26. Yang, B., Liu, D.: Force-based incremental algorithm for mining community structure in dynamic network. J. Comput. Sci. Technol. **21**, 393–400 (2006)
27. Cazabet, R., Rossetti, G.: Challenges in community discovery on temporal networks. In: Holme, P., Saramäki, J. (eds.) Temporal Network Theory. CSS, pp. 185–202. Springer, Cham (2019). https://doi.org/10.1007/978-3-030-23495-9_10
28. Mohammadmosaferi, K.K., Naderi, H.: Evolution of communities in dynamic social networks: an efficient map-based approach. Expert Syst. Appl. **147**, 113221 (2020)
29. Zarayeneh, N., Kalyanaraman, A.: A fast and efficient incremental approach toward dynamic community detection. In: 2019 IEEE/ACM International Conference on Advances in Social Networks Analysis and Mining (ASONAM), Vancouver, pp. 9–16. IEEE (2019). https://doi.org/10.1145/3341161.3342877
30. Kipf, T., Welling, M.: Semi-supervised classification with graph convolutional networks. ArXiv (2016)
31. Long, M., Johnson, D.D.: Graph convolutional networks for node classification and regression tasks. Neural Netw. (2018)
32. Hochreiter, S., Schmidhuber, J.: Long short-term memory. Neural Comput. **9**(8), 1735–1780 (1997)
33. Chung, J., Gülçehre, Ç., Cho, K.H., Bengio, Y.: Empirical evaluation of gated recurrent neural networks on sequence modeling. ArXiv (2014)
34. Huang, X., et al.: DGraph: a large-scale financial dataset for graph anomaly detection. In: Advances in Neural Information Processing Systems, vol. 35, pp. 22765–22777 (2022)
35. Newman, M.E.J., Leicht, E.A.: Detecting community structure in networks. Proc. Natl. Acad. Sci. **104**(2), 836–841 (2007)
36. Jain, A.K.: Data clustering: 50 years beyond k-means. Pattern Recogn. Lett. **31**, 651–666 (2008)
37. Blondel, V.D., Guillaume, J., Lambiotte, R., Lefebvre, E.: Fast unfolding of communities in large networks. J. Stat. Mech: Theory Exp. **2008**, P10008 (2008)

38. Raghavan, U.N., Albert, R., Kumara, S.R.: Near linear time algorithm to detect community structures in large-scale networks. Phys. Rev. E Stat. Nonlinear Soft Matter Phys. **76**(3), 036106 (2007)
39. Liu, F., Wu, J., Xue, S., Zhou, C., Yang, J., Sheng, Q.Z.: Detecting the evolving community structure in dynamic social networks. World Wide Web **23**, 715–733 (2019)
40. Appel, A.P., Cunha, R.L.F., Aggarwal, C.C., Terakado, M.M.: Temporally evolving community detection and prediction in content-centric networks. In: Berlingerio, M., Bonchi, F., Gärtner, T., Hurley, N., Ifrim, G. (eds.) ECML PKDD 2018. LNCS, vol. 11052, pp. 3–18. Springer, Cham (2019). https://doi.org/10.1007/978-3-030-10928-8_1

From Tweets to Token Sales: Assessing ICO Success Through Social Media Sentiments

Donghao Huang[ID], Samuel Samuel[ID], Quoc Toan Hyunh[ID],
and Zhaoxia Wang[✉][ID]

School of Computing and Information Systems, Singapore Management University,
80 Stamford Rd, Singapore 178902, Singapore
{dh.huang.2023,samuel1.2021,qthuynh.2020,zxwang}@smu.edu.sg

Abstract. With the advent of social network technology, the influence of collective opinions has significantly impacted business, marketing, and fundraising. Particularly in the blockchain space, Initial Coin Offerings (ICOs) gain substantial exposure across various online platforms. Yet, the intricate relationships among these elements remain largely unexplored. This study aims to investigate the relationships between social media sentiment, engagement metrics, and ICO success. We hypothesize a positive correlation between favorable sentiment in ICO-related tweets and overall project success. Additionally, we recognize social media engagement indicators (mentions, retweets, likes, follower counts) as critical factors affecting ICO performance. Employing machine learning techniques, we conduct sentiment analysis on tweets, discerning emotional nuances and categorizing expressions as positive or negative. Employing established classification methods, we further analyze engagement data to reveal its impact on ICO interest and awareness. Our research findings offer insights into the predictive potential of social media strategies for ICO success and underscore the importance of investor sentiment and engagement in the volatile cryptocurrency landscape. These insights provide actionable guidance for aspiring crypto founders in formulating effective business development strategies.

The source codes and datasets of this paper are accessible at GitHub: https://github.com/inflaton/Success-Indicators-of-Initial-Coin-Offeri ngs.

Keywords: Cryptocurrency · Initial Coin Offerings (ICOs) ·
Sentiment analysis · Social media · Machine Learning

1 Introduction

The emergence of blockchain technology and cryptocurrencies has transformed fundraising through methods like Initial Coin Offerings (ICOs) and Initial Exchange Offerings (IEOs) [9]. ICOs, also known as token sales, provide cost-effective fundraising avenues by issuing new coins on blockchain-based platforms. Social media platforms, particularly Twitter, have significantly contributed to the promotion and dissemination of information about both ICOs and IEOs [27].

Z. Wang and C. W. Tan (Eds.): PAKDD 2024 Workshops, LNAI 14658, pp. 57–69, 2024.
https://doi.org/10.1007/978-981-97-2650-9_5

IEOs, facilitated by cryptocurrency exchanges, involve these platforms managing fundraising campaigns by issuing new tokens [10]. Acting as intermediaries between projects and investors, these exchanges offer comprehensive crowd sale services [28]. The ICO model witnessed substantial funding, with 3,782 ICOs launched in 2018 alone, collectively raising nearly $11.4 billion [6,17]. Despite fluctuations, the prevalence of ICOs persisted in 2021, with 113 hosted on Coincodex [13,17]. The foundation of an ICO rests on its whitepaper, outlining project details such as business strategy, token sale structure, fund allocation, team composition, and development roadmap [13]. Utilizing the broad reach of social media, ICOs frequently leverage platforms like Twitter to promote their offerings within the crypto community [14].

Businesses utilize Airdrop initiatives to increase awareness about emerging ICOs or IEOs, rewarding participants with digital tokens or cryptocurrency for engaging in marketing activities. This involves actions like joining Telegram groups, following on social media platforms, retweeting, and tagging friends, fostering community engagement and brand endorsement [18]. Social media, especially Twitter, plays a vital role in the success of ICOs and IEOs, offering swift, cost-effective means to raise capital, supported by blockchain's efficiency in creating digital tokens through smart contracts, ensuring security and reliability [25].

Existing studies on ICOs and IEOs often focus solely on direct social media mentions, neglecting the potential impact of indirect interactions. This study aims to bridge this gap by comprehensively analyzing both direct and indirect social media interactions related to crypto projects. It explores sentiment and engagement levels in direct tweets, replies mentioning token keywords, and project announcements on Twitter. By predicting ICOs success using social media sentiment and engagement, this study seeks to uncover their interrelation, offering insights into the factors influencing ICOs and IEOs success on social media.

The main contributions of this paper are summarized as follows:

1. The study delves into the intricate relationships among social media, such as Tweets, and ICO success in the blockchain space, which remains largely unexplored. This contributes to filling the gap in understanding how these elements interact and influence each other.
2. By employing machine learning techniques for social media sentiment analysis and analyzing engagement data, the research aims to reveal the predictive potential of social media strategies for ICO success. This contribution offers valuable insights for crypto founders, indicating the importance of effective social media utilization in driving project success.
3. The research findings offer actionable guidance for aspiring crypto founders by emphasizing the significance of investor sentiment and engagement in the volatile cryptocurrency landscape. This contribution provides practical advice for formulating effective business development strategies in the blockchain industry.

2   Related Work and Motivations

ICOs have become a popular means for startups to raise funds. However, the success of ICOs is influenced by various factors that need to be considered for a comprehensive understanding of their fundraising outcomes.

As social media platforms continue to evolve, there has been a notable surge in individual and organizational engagement, with a growing emphasis on expressing and sharing opinions [11,16]. Consequently, social media sentiment analysis has garnered widespread popularity [26,30,33]. Such sentiment analysis of social media serves as a vital method for individuals and organizers facilitating a deeper understanding of prevailing opinions and sentiments [20,31,32,34], thereby aiding decision-making processes crucial for business success.

Social media sentiment analysis has become indispensable in the financial domain, supported by numerous studies highlighting its effectiveness and contributions [19,29,31]. While these studies have identified various predictors of success, Campino et al. stress the necessity for a more comprehensive analysis of social media's evolving role in ICO outcomes [8]. Despite identifying key predictors such as third-party ratings and detailed whitepapers, the study lacks depth in exploring the multifaceted influence of social media on ICO performance.

Some researchers have underscored the importance of factors such as detailed project information, idea uniqueness, and team competencies [3]. Yet, they often overlook the significant influence of social media on investor behavior. Another study identified conditions to enhance ICO credibility, including registration, GitHub code publication, and early funding [5]. However, this overlooks potential drawbacks such as the effort required to secure funding and risks associated with code transparency.

In a separate analysis, Lyandres et al. identified various ICO success factors, including hardcap, whitepaper informativeness, pre-ICO social media activity, bonuses, and transparency [22]. However, its limited time scope and lack of consideration for negative sentiment impact are notable limitations.

Twitter sentiment analysis plays a pivotal role in evaluating ICO effectiveness [2]. The platform's global reach facilitates succinct communication, fostering high engagement rates [24]. Analyzing meticulously collected Twitter comments related to ICOs, sentiment analysis unveils behavioral and social sentiment patterns among investors. These insights serve as a foundation for devising effective marketing strategies [7].

Influencer-driven product tweets significantly influence brand trust and consumer decisions. Twitter discussions bolster brand recognition and attract new audiences, leading to increased sales [4]. Our study aims to comprehensively explore global social media strategies' impact on ICO success, contrasting with regional studies [12]. Understanding social media's influence aids startups in refining marketing strategies and engaging with investors. Analyzing social media metrics and sentiment guides effective investor targeting and platform selection for ICO promotion. A holistic social media study offers invaluable insights for ICO fundraising efforts.

3   Hypotheses

Hypothesis 1: Positive sentiment expressed in tweets mentioning an ICO is correlated with its success. Sentiment analysis of tweets can provide valuable insights into public perception of a cryptocurrency project. The NRC Word-Emotion Association Lexicon can be used to analyse the emotional content of tweets and identify positive and negative sentiments. A higher proportion of positive tweets is likely to indicate a successful ICO. Sentiment analysis can be performed by collecting and analyzing tweets mentioning the ICO and their respective replies. Table 1 illustrates some examples of original and reply tweet data that were used for the analysis.

Table 1. Example of data used for sentiment analysis.

	Tweet Content	Sentiment
Original Tweet	Just bought more $SOL, feeling great about this investment!!	Positive
Reply	I'm not convinced, I think $SOL is overvalued	Negative
Reply	Quite bullish on it too, the fundamentals are really strong	Positive
Reply	Waiting to see how the market develops, but currently looking for better options in the market	Negative

Hypothesis 2: The level of engagement on social media, such as the number of mentions, retweets, likes, and followers, is positively associated with the success of an ICO. Engagement on social media can indicate the level of awareness and interest in a cryptocurrency project among potential investors. By collecting data on the number of mentions, retweets, likes, and followers of the project and the users mentioning it, we can measure the level of engagement and use it to predict the success of an ICO. Table 2 illustrates an example of the engagement metrics used, and the number of each metric collected for the $MATIC crypto token.

Table 2. Example of data used for engagement metrics.

Metric	Data collected for $MATIC
Number of mentions	799,849
Number of retweets	1,471,118
Number of likes	1,280,414
Number of project followers	34,755,686

4   Dataset

ICOs data was collected from ICOBench[1] and CryptoRank[2] using a combination of manual download and automatic scraping using Python beautiful soup package[3]. For each cryptocurrency project, we recorded ICO-related data as summarized in Table 3.

Table 3. ICO data collected.

Data description	Data type	Hypothesis
Project name	String	1, 2
Project token	String	1, 2
Soft cap	Integer	1, 2
Total raised	Integer	1, 2
ICO Date	Date	1, 2

Data from Twitter was gathered using the Twitter API and Scrapy[4], an open-source Python framework. The collection encompassed two categories of tweets: those that included a cryptocurrency project's keyword, marked by the prefix "$" followed by the token's name (e.g., "$SOL" or "$MATIC"), and tweets emanating from the project's official Twitter handle. For tweets featuring the token keyword, we compiled data on the number of retweets, likes, and the poster's follower count, in addition to the content of any responses. However, tweets originating from the project's own account were excluded from direct analysis as they primarily represent official communications from the cryptocurrency project's team, thus deemed not directly pertinent to the research hypotheses.

Tweets are categorized into direct mentions and indirect mentions. A direct mention occurs when a tweet explicitly contains the token keyword. An indirect mention refers to a tweet that either responds to another tweet featuring the token keyword or replies to a tweet from the project's official Twitter handle. It's important to highlight that to avoid look-ahead bias, only tweets published prior to the dates of a token's ICO are considered relevant for assessing the ICO's prospective success.

After collecting the above data, we processed and transformed it to obtain useful information. For each cryptocurrency project, we recorded the following twitter data as summarized in Table 4:

---

[1] www.icobench.com.

[2] www.cryptorank.io.

[3] https://beautiful-soup-4.readthedocs.io/.

[4] https://scrapy.org/.

Table 4. Twitter data collected.

Data description	Data type	Hypothesis
Content of tweets - direct mentions	List of String	1
Content of tweets - indirect mentions	List of String	1
Total number of direct mentions	Integer	2
Total number of indirect mentions	Integer	2
Total number of likes	Integer	2
Total number of retweets	Integer	2
Number of official account followers	Integer	2
Total number of followers of users mentioning project	Integer	2

5   Experiment and Methodology

5.1   Data Processing

For the facilitation of machine learning algorithms in the examination of our hypotheses, data preprocessing was undertaken to convert the raw data into a format amenable to algorithmic training.

Sentiment analysis was a pivotal component of our study, aimed at capturing the emotional undercurrents embedded in both overt and covert mentions. This was achieved by leveraging the NRC Word-Emotion Association Lexicon[5], a compendium that assigns sentiments-either positive or negative-to various words. We utilized TextBlob[6], a Python library recognized for its simple API that enables execution of several natural language processing (NLP) tasks, including sentiment analysis. This was instrumental in quantifying the sentiment polarity of each tweet. Figure 1 provides a visual representation of the sentiment evaluation conducted on tweets pertaining to the DUO Network (DUO) ICO.

The success of an ICO was quantitatively determined by comparing the total capital raised against the soft cap defined by the project, with the latter serving as the minimum required funding level. In our corpus of 816 ICOs, 582 were classified as having achieved success by surpassing their respective soft caps, whereas 234 were deemed unsuccessful for not meeting these financial benchmarks. The dataset was structured into 12 columns, each representing a discrete attribute of an ICO from a specific cryptocurrency project. A condensed overview of these attributes is delineated in Table 8, which can be found in the Appendix.

5.2   Machine Learning Methods

The objective of the current research was to forecast the potential success of Initial Coin Offerings (ICOs) by categorizing them as likely to succeed or not.

---

[5] https://saifmohammad.com/WebPages/NRC-Emotion-Lexicon.htm.
[6] https://textblob.readthedocs.io/.

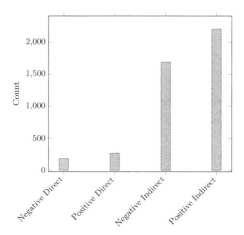

Fig. 1. Sentiment scores for DUO Network (DUO) tweets.

This was based on an array of input variables, all represented by integer values, including attributes like 'soft_cap', 'total_positive_direct_mentions', among others. We segmented the data into two distinct groups: one for training, constituting 80% of the entire dataset, and the other for testing, making up the remaining 20%. Our methodological approach encompassed the deployment of six diverse algorithms, specifically Support Vector Machines (SVMs), Logistic Regression, Random Forest, Naïve Bayes, Categorical Boosting (CatBoost), and Neural Network. These methods were selected for their proven competency in executing classification tasks. The process of parameter optimization was conducted through grid search with the aim of maximizing accuracy.

6 Results and Discussion

6.1 Individual Hypothesis Testing

The classification methods were first applied to hypothesis 1 - "Positive sentiment expressed in tweets mentioning an ICO is correlated with its success" by using only the x variables directly linked to sentiments (variables 2, 3, 5, 6 in Table 8). Table 5 illustrates the performance outcomes of each predictive model.

Next, the same methods were applied to hypothesis 2 - "The level of engagement on social media, such as the number of mentions, retweets, likes, and followers, is positively associated with the success of an ICO" by using the remaining variables which are directly linked to level of engagement. Table 6 presents the performance outcomes for each predictive model.

The results shown in Table 5 suggest that the models trained to test Hypothesis 1 yielded moderate performance scores, with the highest accuracy achieved by the Neural Network model at 74.4%. The F1 Scores ranged from 70.9% to 81.9%, which is moderately high. These results indicate that Hypothesis 1 has

Table 5. Hypothesis 1 predictive results

Model	Accuracy	Precision	Recall	F1 Score
Naïve Bayes	65.9%	66.0%	98.1%	78.9%
SVM	65.2%	65.2%	**100.0%**	79.0%
Logistic Regression	66.5%	**81.7%**	62.6%	70.9%
Random Forest	73.8%	75.4%	88.8%	81.5%
CatBoost	73.8%	75.4%	88.8%	81.5%
Neural Network	**74.4%**	76.0%	88.8%	**81.9%**

Table 6. Hypothesis 2 predictive results

Model	Accuracy	Precision	Recall	F1
Naïve Bayes	64.6%	65.4%	97.2%	78.2%
SVM	65.2%	65.2%	**100.0%**	79.0%
Logistic Regression	65.2%	**82.9%**	58.9%	68.9%
Random Forest	**76.8%**	78.0%	89.7%	**83.5%**
CatBoost	76.2%	76.6%	91.6%	83.4%
Neural Network	73.8%	76.2%	86.9%	81.2%

some merit and performs decently in predicting the success of ICOs based on positive sentiment expressed in tweets.

In contrast, the models trained to test Hypothesis 2 in Table 6 performed better overall, with the highest accuracy achieved by the Random Forest model at 76.8%. The Precision and Recall scores for all models were relatively high, ranging from 65.2% to 82.9% for Precision and 58.9% to 100.0% for Recall, respectively. The F1 Scores ranged from 68.9% to 83.5%, which is slightly higher than the accuracy/F1 scores achieved in the models trained to test Hypothesis 1. These results suggest that while sentiment expressed through tweets may be a moderately strong predictor of ICO success (hypothesis 1), the level of engagement a crypto project has on social media might act as a better predictor (hypothesis 2), supporting the plausibility of hypothesis 2 over hypothesis 1.

6.2   Combined Hypotheses Testing

A combination of the feature sets from both hypotheses was tested for predictive power. This appeared to improve the overall performance, particularly for the CatBoost model, as seen in Table 7.

The findings indicate that both Hypothesis 1 and Hypothesis 2 could be effective in forecasting ICO success, particularly when applied in conjunction. Among the tested models, CatBoost and Random Forest emerged as the top performers in accuracy.

Table 7. Combined predictive results

Model	Accuracy	Precision	Recall	F1
Naïve Bayes	64.0%	65.2%	96.3%	77.7%
SVM	65.2%	65.2%	**100.0%**	79.0%
Logistic Regression	64.6%	85.5%	55.1%	67.0%
Random Forest	**78.0%**	**79.3%**	89.7%	84.2%
CatBoost	**78.0%**	77.1%	94.4%	**84.9%**
Neural Network	67.1%	66.5%	**100%**	79.9%

6.3   Further Discussion

There are several potential reasons why the level of engagement on social media may be a more reliable predictor of the success of an ICO than positive sentiment expressed in tweets. One reason is that the level of engagement on social media is a more comprehensive measure of interest and support for a project, considering factors such as the number of mentions, retweets, likes, and followers [1]. Positive sentiment expressed in tweets, on the other hand, may not capture all forms of engagement and may be influenced by factors such as bots and fake accounts [23].

Furthermore, the sentiment expressed in tweets may not always reflect the true feelings and opinions of the wider community [21]. For example, some people may not reply to tweets or publicly express their support or interest in a project, but may still be engaged with the project on other levels, such as following its social media pages or participating in ICOs and purchasing the tokens without posting anything on Twitter. In contrast, the level of engagement on social media provides a more comprehensive and holistic view of the level of interest and engagement in a project.

Another potential reason why positive sentiment expressed in tweets may not be as strong an indicator of success as the level of engagement on social media is that sentiment can be influenced by a variety of factors, including competitors, influencers, and external events. For example, a project that is in direct competition with another project may receive a high volume of negative sentiment simply because of the rivalry, even if the project itself is strong and has potential for success. Therefore, sentiment expressed in tweets may not always accurately reflect the actual quality and potential of a project.

Another point to consider is that negative sentiment expressed on social media can sometimes have unintended positive effects. In the age of information overload, negative tweets or comments may grab more attention than positive ones and can generate buzz and discussion around a project, leading to increased visibility and exposure [15]. It is also important to note that bad marketing is still marketing, and negative sentiment can still create awareness and interest in a project. Therefore, when evaluating the potential success of a project, it is essential to consider both positive and negative sentiment expressed on social media. While positive sentiment is a good indicator of potential success, it should not be the sole metric to evaluate a project, as it can be influenced by external factors such as competitors or influencers. Ultimately, a more comprehensive view of sentiment and engagement on social media is necessary to accurately evaluate the potential success of a project.

Overall, the level of engagement on social media appears to be a more reliable predictor of the success of an ICO than positive sentiment expressed in tweets. The comprehensive and holistic nature of engagement metrics such as mentions, retweets, likes, and followers, combined with their relative independence from external factors such as competition and sentiment, make them a more robust and reliable measure of interest and support for a project.

Our findings highlight the crucial role of social media engagement in predicting ICO success, suggesting a strategic focus on enhancing online interactions. ICO projects should prioritize increasing social media activities like mentions, retweets, and followers to attract and sustain interest. A concise strategy would involve creating engaging content, managing active communities, and collaborating with influencers to amplify visibility. This approach underscores the importance of both engagement quantity and sentiment quality, guiding projects to optimize their social media presence for better funding outcomes.

7   Conclusion

This research explores how Twitter sentiment and engagement influence ICO fundraising outcomes. It was discovered that a positive sentiment on Twitter correlates with greater fundraising achievements. Twitter serves as a crucial channel for ICO promotion and a source for investors to obtain cryptocurrency-related information. The findings indicate that ICOs receiving more positive commentary and engagement on Twitter tend to secure higher funding amounts. Notably, a strong relationship was observed between the fundraising success of an ICO and both its Twitter sentiment and the volume of its direct and indirect followers. To assess predictive accuracy, six classifiers were employed: Support Vector Machines, Logistic Regression, Random Forest, Naïve Bayes, Categorical Boosting, and Neural Network. Among these, Random Forest and Categorical Boosting emerged as the most effective, leading in performance across two of the four evaluation metrics.

In addition to the current research, there are still many avenues to explore regarding the impact of social media on ICO success. One potential area of future work is to investigate "pump and dump" ICO projects, where the followers and their respective interactions are likely to be AI bots and paid Twitter accounts, specially paid to pump the projects. These types of projects have been known to manipulate social media sentiment and create an artificial hype around the project, leading to short-term gains but ultimately resulting in significant losses for unsuspecting investors. Understanding the impact of these tactics on ICO success could be valuable in detecting and avoiding such fraudulent projects.

Another area for future research is to explore how a few extremely influential figures can single-handedly affect the price of a crypto project. The impact of these influential figures on social media sentiment can be significant and may lead to short-term gains or losses for investors. Studying the impact of such influencers on ICO success could provide insights into the role of individual powerhouses in the cryptocurrency market and could lead to a better understanding of how to mitigate risks associated with their influence.

Additionally, further research is needed to explore the connection between social media marketing expenses and the financial performance of ICO projects. Investigating how an ICO's governance signals influence fundraising success can provide valuable insights into key success factors. Furthermore, understanding the broader impact of cryptocurrencies and blockchain technologies on businesses is crucial for predicting the future of ICOs.

Acknowledgment. The authors extend their sincere appreciation to the SMU student assistants, namely Jolie Zhi Yi FONG, and Yvonne LIM, for their contributions to this research. The authors would like to thank Prof. Zhiping LIN and Juncheng CHEN, an NTU PhD student, for the valuable discussions.

A Appendix

Table 8. Feature labels and description.

Index	Column	Data type	Description
1	total_direct_mentions	Integer	Total number of direct mentions
2	total_positive_direct_mentions	Integer	Total number of direct mentions with positive sentiment
3	total_negative_direct_mentions	Integer	Total number of direct mentions with negative sentiment
4	total_indirect_mentions	Integer	Total number of indirect mentions
5	total_positive_indirect_mentions	Integer	Total number of indirect mentions with positive sentiment
6	total_negative_indirect_mentions	Integer	Total number of indirect mentions with negative sentiment
7	total_retweets	Integer	Total retweets of tweets with token keyword and official account tweets
8	total_likes	Integer	Total likes of tweets with token keyword and official account tweets
9	total_project_followers	Integer	Total followers of official account
10	total_indirect_followers	Integer	Total followers of users mentioning the project
11	soft_cap	Integer	Project soft cap
12	ico_success	Integer	Boolean (0 or 1) to indicate success of ICO

References

1. Albrecht, S., Lutz, B., Neumann, D.: How sentiment impacts the success of blockchain startups–an analysis of social media data and initial coin offerings (2019)
2. Albrecht, S., Lutz, B., Neumann, D.: The behavior of blockchain ventures on Twitter as a determinant for funding success. Electron. Mark. **30**(2), 241–257 (2020)
3. Alchykava, M., Yakushkina, T.: ICO performance: analysis of success factors. In: IMS, pp. 241–249 (2021)
4. Appel, G., Grewal, L., Hadi, R., Stephen, A.T.: The future of social media in marketing. J. Acad. Mark. Sci. **48**(1), 79–95 (2020)
5. Belitski, M., Boreiko, D.: Success factors of initial coin offerings. J. Technol. Transf. **47**(6), 1690–1706 (2022)
6. Boreiko, D., Vidusso, G.: New blockchain intermediaries: do ICO rating websites do their job well? (2019)
7. Calderwood, L.U., Soshkin, M.: The travel and tourism competitiveness report 2019. World Economic Forum (2019)
8. Campino, J., Brochado, A., Rosa, A.: Success factors of initial coin offering (ICO) projects. Econ. Bull. **41**(2), 252–262 (2021)
9. Campino, J., Brochado, A., Rosa, Á.: Initial coin offerings (ICOs): the importance of human capital. J. Bus. Econ. **91**, 1225–1262 (2021)
10. Chamorro Domínguez, M.C.: Financing of start-ups via initial coin offerings and gender equality. In: Miller, K., Wendt, K. (eds.) The Fourth Industrial Revolution and Its Impact on Ethics, pp. 183–197. Springer, Cham (2021). https://doi.org/10.1007/978-3-030-57020-0_14
11. Chen, Z., Wang, Z., Lin, Z., Yang, T.: Comparing ELM with SVM in the field of sentiment classification of social media text data. In: Cao, J., Vong, C., Miche, Y., Lendasse, A. (eds.) ELM 2018. PALO, vol. 11, pp. 336–344. Springer, Cham (2020). https://doi.org/10.1007/978-3-030-23307-5_36
12. Chursook, A., Dawod, A.Y., Chanaim, S., Naktnasukanjn, N., Chakpitak, N.: Twitter sentiment analysis and expert ratings of initial coin offering fundraising: evidence from Australia and Singapore markets. TEM J. **11**(1), 44 (2022)
13. Daskalakis, N., Georgitseas, P.: An Introduction to Cryptocurrencies: The Crypto Market Ecosystem, pp. 41–55. Routledge, London (2020)
14. Drobetz, W., Momtaz, P.P., Schröder, H.: Investor sentiment and initial coin offerings. J. Altern. Investments **21**(4), 41–55 (2019)
15. Freeman, D., McWilliams, T., Bhattacharyya, S., Hall, C., Peillard, P.: Enhancing trust in the cryptocurrency marketplace: a reputation scoring approach. SMU Data Sci. Rev. **1**(3), 5 (2018)
16. Fu, X., et al.: Social media for supply chain risk management. In: 2013 IEEE International Conference on Industrial Engineering and Engineering Management, pp. 206–210. IEEE (2013)
17. Giudici, G., Adhami, S.: The impact of governance signals on ICO fundraising success. J. Ind. Bus. Econ. **46**(2), 283–312 (2019)
18. Howell, S.T., Niessner, M., Yermack, D.: Initial coin offerings: financing growth with cryptocurrency token sales. Rev. Financ. Stud. **33**(9), 3925–3974 (2020)
19. Hu, Z., Wang, Z., Ho, S.B., Tan, A.H.: Stock market trend forecasting based on multiple textual features: a deep learning method. In: 2021 IEEE 33rd International Conference on Tools with Artificial Intelligence (ICTAI), pp. 1002–1007. IEEE (2021)

20. Hu, Z., Wang, Z., Wang, Y., Tan, A.H.: MSRL-Net: a multi-level semantic relation-enhanced learning network for aspect-based sentiment analysis. Expert Syst. Appl. **217**, 119492 (2023)

21. Kummer, S., Herold, D.M., Dobrovnik, M., Mikl, J., Schäfer, N.: A systematic review of blockchain literature in logistics and supply chain management: identifying research questions and future directions. Future Internet **12**(3), 60 (2020). https://doi.org/10.3390/fi12030060. https://www.mdpi.com/1999-5903/12/3/60

22. Lyandres, E., Palazzo, B., Rabetti, D.: ICO success and post-ICO performance (2020)

23. Mirtaheri, M., Abu-El-Haija, S., Morstatter, F., Ver Steeg, G., Galstyan, A.: Identifying and analyzing cryptocurrency manipulations in social media. IEEE Trans. Comput. Soc. Syst. **8**(3), 607–617 (2021)

24. Mohd-Sulaiman, A.N., Hingun, M.: Liability risks in shareholders' engagement via electronic communication and social media. Int. J. Law Manag. **62**(6), 539–555 (2020)

25. Morkunas, V.J., Paschen, J., Boon, E.: How blockchain technologies impact your business model. Bus. Horiz. **62**(3), 295–306 (2019)

26. Teo, A., Wang, Z., Pen, H., Subagdja, B., Ho, S.B., Quek, B.K.: Knowledge graph enhanced aspect-based sentiment analysis incorporating external knowledge. In: 2023 IEEE International Conference on Data Mining Workshops (ICDMW), pp. 791–798. IEEE (2023)

27. Tiwari, M., Gepp, A., Kumar, K.: The future of raising finance-a new opportunity to commit fraud: a review of initial coin offering (ICOs) scams. Crime Law Soc. Change **73**, 417–441 (2020)

28. Vivion, M., Hennequin, C., Verger, P., Dubé, E.: Supporting informed decision-making about vaccination: an analysis of two official websites. Public Health **178**, 112–119 (2020)

29. Wang, Z., Ho, S.B., Lin, Z.: Stock market prediction analysis by incorporating social and news opinion and sentiment. In: 2018 IEEE International Conference on Data Mining Workshops (ICDMW), pp. 1375–1380. IEEE (2018)

30. Wang, Z., Hu, Z., Ho, S.B., Cambria, E., Tan, A.H.: MiMuSA–mimicking human language understanding for fine-grained multi-class sentiment analysis. Neural Comput. Appl. **35**, 15907–15921 (2023)

31. Wang, Z., Hu, Z., Li, F., Ho, S.B., Cambria, E.: Learning-based stock trending prediction by incorporating technical indicators and social media sentiment. Cogn. Comput. **15**(3), 1092–1102 (2023)

32. Wang, Z., Tong, V.J.C., Chan, D.: Issues of social data analytics with a new method for sentiment analysis of social media data. In: 2014 IEEE 6th International Conference on Cloud Computing Technology and Science, pp. 899–904. IEEE (2014)

33. Wang, Z., Tong, V.J.C., Chin, H.C.: Enhancing machine-learning methods for sentiment classification of web data. In: Jaafar, A., et al. (eds.) AIRS 2014. LNCS, vol. 8870, pp. 394–405. Springer, Cham (2014). https://doi.org/10.1007/978-3-319-12844-3_34

34. Wang, Z., Tong, V.J.C., Xin, X., Chin, H.C.: Anomaly detection through enhanced sentiment analysis on social media data. In: 2014 IEEE 6th International Conference on Cloud Computing Technology and Science, pp. 917–922. IEEE (2014)

Enhanced Graph Neural Network for Session-Based Recommendation with Static and Dynamic Information

Yongxin Chao[ID] and Kai Zheng[(✉)] [ID]

School of Data Science and Engineering, East China Normal University, Shanghai, China
51215903090@stu.ecnu.edu.cn, kzheng@cs.ecnu.edu.cn

Abstract. Session-based recommendation (SBR) is a complex endeavor focused on predicting a user's next interesting item based on his sessions (i.e., short interaction sequences). The existing SBR models usually learn only one aspect of the sessions, either the static information (e.g., spatial structure of the graph, node similarity) or the dynamic information (e.g., temporal information, position information), so that the rich information embedded in session can't be fully exploited. A new enhanced graph neural network model based on both static and dynamic information (called EGNN-SDI) is proposed, which constructs and uses a global graph that cooperates with an undirected and a directed session graph to learn the global and local static information, as well as the dynamic information within sessions. Based on this, we propose a new node encoding layer called SDI.

In short, we use GCN and GGNN separately to learn static and dynamic information, respectively. An inverse position matrix is also introduced to learn the relative positional information within the session. By using linear combination and attention mechanisms, the enhanced item representation enables the generation of more accurate session representations for the next item to be recommended. Evaluation experiments are performed on three widely used datasets, consistently showcasing the superiority of EGNN-SDI compared to existing baseline models. Our model's implementation can be accessed via https://anonymous.4open.sci ence/r/EGNN-SDI-4B8C.

Keywords: Session-based recommendation · Graph neural network · Attention mechanism

1 Introduction

With the continuous expansion of the Internet and the exponential growth of digital content, users face the problem of information overload. Recommender systems, crucial for addressing this challenge, facilitating users in discovering and acquiring content tailored to their interests. Traditional recommendation algorithms, like collaborative filtering, primarily depend on long-term historical interactions and user data, which often lead to sub-optimal results on popular platforms such as streaming media and e-commerce (e.g., TikTok, YouTube and Taobao). Therefore, SBR has been proposed

© The Author(s), under exclusive license to Springer Nature Singapore Pte Ltd. 2024
Z. Wang and C. W. Tan (Eds.): PAKDD 2024 Workshops, LNAI 14658, pp. 70–81, 2024.
https://doi.org/10.1007/978-981-97-2650-9_6

as a novel paradigm in recommender systems. It is designed to capture users' dynamic, short-term preferences, thereby facilitating the provision of more timely and accurate recommendations.

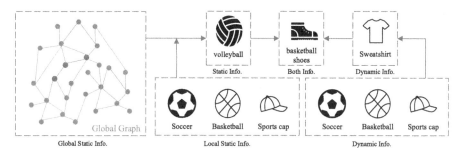

Fig. 1. A simple example of the role of static and dynamic information on SBR.

The Markov Chain (MC) model [6, 13] has been applied to SBR with notable success. However, these models have a primary limitation in that they focus exclusively on the most recent item within the ongoing session, assuming significant independence between items. To improve model performance, the Recurrent Neural Network (RNN) has been widely used in session-based recommendation [2–5] for its ability to capture longer interaction sequences. By exploiting the temporal continuity of the data, the RNN demonstrates superior performance in various session-based recommendation scenarios. While significant performance has been achieved, these methods may not adequately capture the user's interest and tend to overlook complex item transformations. Recent work uses graph structures to represent sessions in order to encapsulate richer information and utilize Graph Neural Networks (GNNs) for information propagation among neighboring items, leading to enhanced performance. However, most graph GNN-based methods primarily focus on learning either the static information of sessions (e.g., the spatial structure of the graph, node similarity) or the dynamic aspects (such as timing and location information). For instance, GCE-GNN [7] acquires the spatial structure of the global session graph through an attention mechanism, while SR-GNN [8] captures temporal information within sessions via the Gated Graph Neural Network (GGNN). Simultaneously incorporating both static and dynamic session information can provide the model with a more comprehensive understanding and utilization of session data, ultimately enhancing recommendation performance.

As shown in Fig. 1, for a given session, when the model learns only static information, it tends to recommend items closely aligned with the overarching theme of the session, such as volleyball. This is because the session has the most ball items. On the other hand, if the model only focuses on dynamic information, it will place more emphasis on the user's behavior and preferences within the current session, leading to recommendations for other types of sports equipment or related sports products, such as sportswear. However, a model that integrates both static and dynamic information can achieve a balance between global preferences and personalized user behavior. Such a model is likely to recommend sports equipment related to the user's interest in ball sports, e.g., suggesting basketball shoes.

The paper proposes a novel model that learns two types of information simultaneously, called Enhanced Graph Neural Network based on static and dynamic session information (EGNN-SDI). In EGNN-SDI, we create a global graph that includes all available data, as well as undirected and directed graphs based on the current session. This approach allows us to independently capture both the global and local static attributes of sessions, as well as their dynamic information. As a result, EGNN-SDI not only captures rich session information, but also mitigates the problem of overfitting.

This work's primary contributions are as follows:

- Proposed an improved graph neural network model using an innovative node encoding layer SDI, which effectively captures both static and dynamic information within sessions across the three constructed graphs.
- Introduced a learnable position embedding module and self-attention mechanism into the model, which further improves the learning effect of session sequences.

2 Related Work

Some traditional methods can be repurposed for SBR. K-Nearest Neighbors (KNN) based methods [11, 12] compare the current session with previous ones based on different similarity measures. SKNN [11] combines session context information with the KNN algorithm to compute the similarity between sessions. However, KNN algorithms rely heavily on long-term user interactions and may not capture dynamic changes in user interests. Markov chain (MC) based models [6, 13] predict the next likely clicked item by calculating transition probabilities between items. For example, FPMC [13] combines user-item matrix factorization techniques with MC for next-basket recommendations. However, MC-based methods often struggle to capture more complex sequential patterns.

Due to its powerful ability to model sequence information, RNN has been widely used in the field of SBR [17, 18]. GRU4Rec [2] pioneered RNN-based modeling by employing a multi-layer gated recurrent unit (GRU) to capture the sequence of item interactions. NARM [3] builds upon the foundation of GRU4REC, blending GRU with attention mechanisms. This hybrid encoder is adept at capturing both the user's sequential behavior and their key interests during the current session. SR-IEM [17] delves into users' long-term and recent behavioral patterns to provide personalized recommendations. RCNN-SR [22] combines the GRU and the attention mechanism to understand the general interests of users. Additionally, it employs convolution operations using both horizontal and vertical filters to explore the user's current interests.

In recent years, GNNs have emerged as powerful tools across diverse domains, demonstrating significant promise in representation learning. Moreover, they [15, 20, 21] are also used in the field of SBR. SR-GNN [8] leverages GGNN to model higher-order transitions between items. GCE-GNN [7] combines the pairwise item transformation information of the session graph and the global graph for recommendation. COTREC [9] used a session-based graph to improve both internal and external connections within sessions. It combined self-supervised learning with GNN-based collaborative training to improve model accuracy. HGNN [23] proposed heterogeneous global graphs to model diverse session information. It then used a graph-enhanced hybrid encoder to obtain user preference representations. Disen-GNN [15] is a novel disentangled GNN model that applies decoupled representation learning to embed items in multiple blocks to represent different factors.

Despite the notable advancements achieved by these methods, they primarily focus on either static or dynamic information from sessions, rather than effectively integrating both types of information.

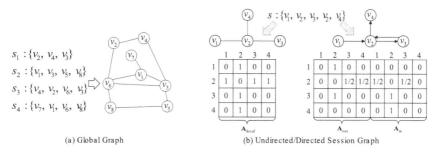

(a) Global Graph   (b) Undirected/Directed Session Graph

Fig. 2. Construction of global graph and session graphs.

3 Preliminary

3.1 Problem Statement

Let $V = \{v_1, v_2, \ldots, v_{|V|}\}$ denotes the unique items that appear in all sessions, where $|V|$ is the number of all unique items. For each anonymous session, it can be denoted by $S = (v_{s,1}, v_{s,2}, \ldots, v_{s,n})$. $V_s \subseteq V$ denote the set of items in current session, where $v_{s,i} \in V_s$ denotes an interaction item within the session S and n denotes the length of the current session. Given a session S, the SBR should predict the next item $v_{s,n+1}$ to be clicked. Therefore, EGNN-SDI takes the session S as input and computes all candidate item probabilities $\hat{y} = \{\hat{y}_1, \hat{y}_2, \ldots, \hat{y}_{|V|}\}$. The items with top-$K$ probabilities are then used for recommendation.

3.2 Graph Construction

Graphs provide an intuitive representation of item interactions within a session, highlighting transitions between adjacent items and potential associations between items. In these graphs, each node represents an item.

First, as shown in Fig. 2(a), we connect all items based on their transition relationships across all sessions, to construct the global graph denoted as $\mathcal{G}_g = \{V, \mathcal{E}\}$. Here, $\mathcal{E} = V \times V$ denote the set of edges, where each edge (v_i, v_j) represents the adjacency of two items. (i.e., they have occurred concurrently in a session.)

As shown in Fig. 2(b), each session can be represented by two types of graphs. Given a session S, the undirected session graph is denoted as $\mathcal{G}_{s,u} = \{V_s, \mathcal{E}_{s,u}\}$. Here, $\mathcal{E}_{s,u}$ denotes the set of undirected edges, where each edge $(v_{s,i}, v_{s,j})$ means that item

$v_{s,i}$ and $v_{s,j}$ are adjacent. The directed session graph is denoted as $\mathcal{G}_{s,d} = \{\mathcal{V}_s, \mathcal{E}_{s,d}\}$. Here, directed edge $(v_{s,i}, v_{s,j}) \in \mathcal{E}_{s,d}$ represent that a user's interaction with item $v_{s,i}$ in a session is immediately followed by an interaction with item $v_{s,j}$. $\mathbf{A}_{in}, \mathbf{A}_{out} \in \mathbb{R}^{n \times n}$ are two normalized weighted connection matrices of graph $\mathcal{G}_{s,d}$.

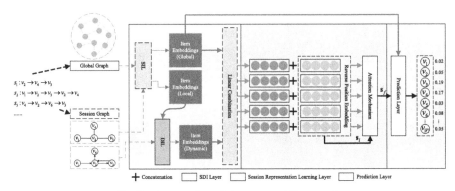

Fig. 3. An overview of the proposed framework.

Each item is embedded into a unified embedding space denoted by $\mathbf{v} \in \mathbb{R}^d$, with d representing the dimension. Similarly, the session S can be denoted by $\mathbf{s} \in \mathbb{R}^d$, which is generated based on the item embeddings.

4 Methodology

The EGNN-SDI architecture is shown in Fig. 3 and mainly consists of three parts: Static and dynamic information enhanced item representation (SDI) learning layer, session representation learning layer and prediction layer.

4.1 Static and Dynamic Information Enhanced Item Representation Learning Layer (SDI)

Static-Driven Item Representation Learning (SIL). The acquisition of static information involves two different facets: one based on the global graph $\mathcal{G}_g$ and the other based on the undirected session graph $\mathcal{G}_{s,u}$. First, we use GCN to capture item transition relationships from a global perspective and to update item embeddings:

$$\Theta^l = \sigma(\mathbf{S}_g \Theta^{l-1} \mathbf{W}_g^{l-1}), \tag{1}$$

where $\mathbf{W}_g^l \in \mathbb{R}^{d \times d}$ is the learnable weight parameter. $\Theta^l \in \mathbb{R}^{|V| \times d}$ denotes the item embeddings after the l-th global GCN layer. $\Theta^0 = \mathbf{X}$ is the initial item representation. $\sigma(\cdot)$ is the sigmoid function. $\mathbf{S}_g = \widetilde{\mathbf{D}}_g^{-\frac{1}{2}} \widetilde{\mathbf{A}}_g \widetilde{\mathbf{D}}_g^{-\frac{1}{2}}$ is the normalized matrix of $\widetilde{\mathbf{A}}_g$. Here, $\widetilde{\mathbf{A}}_g = \mathbf{A}_g + \mathbf{I}_g$ is the adjacency matrix $\mathbf{A}_g$ of the global graph $\mathcal{G}_g$ plus self-connections. And $\mathbf{I}_g$ is the identity matrix. $\widetilde{\mathbf{D}}_g$ is the degree matrix of $\mathbf{A}_g$, where $\widetilde{\mathbf{D}}_{g,ii} = \sum_j \widetilde{\mathbf{A}}_{g,ij}$.

Subsequently, the item representation for the current session S is obtained as $\mathbf{H}_c^0 = \Theta(s)$. To learn consistent feature representations and effectively capture the relationships between global and local information, we also use GCN to capture the local static information of the session S within the undirected session graph. The update function is as follows:

$$\mathbf{H}_c^l = \sigma(\mathbf{S}_c \mathbf{H}_c^{l-1} \mathbf{W}_c^{l-1}), \tag{2}$$

where $\mathbf{W}_c^l \in \mathbb{R}^{d \times d}$ is transformation weight. $\mathbf{H}^l \in \mathbb{R}^{n \times d}$ represents the node embeddings after the l-th layer of local GCN. $\mathbf{S}_c \in \mathbb{R}^{n \times n}$ similar to $\mathbf{S}_g$ is the normalized adjacency matrix of the undirected session graph $\mathcal{G}_{s,u}$.

To enrich the item representation and avoid indistinguishable representations between common nodes in GNN [14], the average pooling technique is used:

$$\mathbf{H}_c = \frac{1}{m} \sum_{l=1}^{m} \mathbf{H}^l, \tag{3}$$

Dynamic-Driven Item Representation Learning (DIL). The static information and dynamic information of the ongoing session are intertwined, and inspired by the success in SR-GNN [8], we use GGNN to comprehensively consider both to update the item representation. For node $v_{s,i} \in \mathcal{G}_{s,d}$, the update function is as follows:

$$\begin{aligned}
\mathbf{a}_i^t &= \text{concat}(\mathbf{A}_{in,i:} \mathbf{V}_s^{t-1} \mathbf{W}_{in} + \mathbf{b}_{in}, \mathbf{A}_{out,i:} \mathbf{V}_s^{t-1} \mathbf{W}_{out} + \mathbf{b}_{out}), \\
\mathbf{z}_i^t &= \sigma(\mathbf{W}_z \mathbf{a}_{s,i}^t + \mathbf{U}_z \mathbf{v}_i^{t-1}), \\
\mathbf{r}_i^t &= \sigma(\mathbf{W}_r \mathbf{a}_i^t + \mathbf{U}_r \mathbf{v}_i^{t-1}), \\
\tilde{\mathbf{v}}_i^t &= \tanh(\mathbf{W}_o \mathbf{a}_i^t + \mathbf{U}_o (\mathbf{r}_i^t \odot \mathbf{v}_i^{t-1})), \\
\mathbf{v}_i^t &= (1 - \mathbf{z}_i^t) \odot \mathbf{v}_i^{t-1} + \mathbf{z}_i^t \odot \tilde{\mathbf{v}}_i^t,
\end{aligned} \tag{4}$$

where $\mathbf{W}_{in}, \mathbf{W}_{out} \in \mathbb{R}^{d \times d}$ are parameter matrices, $\mathbf{b}_{in}, \mathbf{b}_{out} \in \mathbb{R}^d$ are bias vectors. $\mathbf{z}_i^t$ and $\mathbf{r}_i^t$ are the reset and update gates respectively. $\odot$ is the element-wise multiplication operator. $\mathbf{V}_s^t = [\mathbf{v}_1^t, \mathbf{v}_2^t, \ldots, \mathbf{v}_n^t] \in \mathbb{R}^{n \times d}$ represents the item embeddings in session S, and $\mathbf{V}_s^0 = \mathbf{H}_c$. $\mathbf{A}_{in}, \mathbf{A}_{out} \in \mathbb{R}^{n \times n}$ are shown as in Fig. 2(b).

After t layers of GGNN, we get the updated item representation $\mathbf{H}_d = \mathbf{V}^t$. Finally, we derive the representation of each item in the session by combining the global static information with the local session information, and the final representation is obtained by linear combination:

$$\mathbf{H}_f = \alpha * \mathbf{H}_d + \beta * \Theta(s) \tag{5}$$

where α, β are trainable parameters.

4.2 Session Representation Learning Layer

To convert each session into its corresponding embedding vector, we compute the session representation by averaging the item representations $\mathbf{H}_f$:

$$\mathbf{s}_l = \frac{1}{n} \sum_{i=1}^{n} \mathbf{h}_{f,i}, \tag{6}$$

Positional embeddings are an effective technique introduced by Transformer [16]. In user session interaction sequences, positional information is equally important. To preserve the positional information, we concatenate a learnable position-embedding matrix $\mathbf{P} = [\mathbf{p}_1, \mathbf{p}_2, \ldots, \mathbf{p}_n]$ to the item representations captured by SDI, where $\mathbf{p}_i \in \mathbb{R}^d$ represents the positional vector for a specific position i. The positional information is integrated through concatenation and non-linear transformation to obtain item representations that carry positional information:

$$\mathbf{h}_{p,i} = \tanh(\mathbf{W}_p[\mathbf{h}_{f,i} \| \mathbf{p}_{n-i+1}] + \mathbf{b}_p), \tag{7}$$

where $\mathbf{W}_p \in \mathbb{R}^{d \times 2d}$ and $\mathbf{b}_p \in \mathbb{R}^d$ are learnable parameters.

After the extraction by SDI and embedding of positional information, session representations contain complex local information. To optimize the embeddings of session, a soft attention mechanism is used:

$$\begin{aligned} \alpha_i &= \mathbf{q}^\top \sigma(\mathbf{W}_1 \mathbf{s}_l + \mathbf{W}_2 \mathbf{h}_{p,i} + \mathbf{b}), \\ \mathbf{s}_f &= \sum_{i=1}^n \alpha_i \mathbf{h}_{p,i}, \end{aligned} \tag{8}$$

where $\mathbf{q} \in \mathbb{R}^d$ and $\mathbf{W}_1, \mathbf{W}_2 \in \mathbb{R}^{d \times d}$ are the weights controlling the item embeddings.

4.3 Prediction Layer

Utilizing the acquired session representation $\mathbf{s}_f$, we calculate the final recommendation probability for each candidate item by considering its embedding through the global GCN layer and the session representation $\mathbf{s}_f$. This involves performing a dot product followed by a softmax function to generate the output $\hat{\mathbf{y}}$:

$$\hat{\mathbf{y}} = \text{softmax}(\mathbf{s}_f^\top \mathbf{v}_i), \tag{9}$$

Here, $\hat{\mathbf{y}}_i \in \hat{\mathbf{y}}$ represents the probability of item v_i being the next-click in session S.

The loss function is defined as the cross-entropy of the predicted result $\hat{\mathbf{y}}$:

$$\mathcal{L}(\hat{\mathbf{y}}) = -\sum_{i=1}^{|V|} \mathbf{y}_i \log(\hat{\mathbf{y}}_i) + (1 - \mathbf{y}_i) \log(1 - \hat{\mathbf{y}}_i), \tag{10}$$

where $\mathbf{y}$ denotes the one-hot encoding vector of the ground truth item.

5 Experiments

We extensively experiment to assess the effectiveness of the proposed EGNN-SDI method by addressing four crucial research inquiries. RQ1: Does EGNN-SDI outperform SBR baselines on three datasets? RQ2: Does the integration of static and dynamic information effectively improve recommendation performance? RQ3: How does EGNN-SDI perform under different aggregation operations? RQ4: How do different hyperparameters (e.g., number of SIL layer) affect the accuracy of EGNN-SDI?

5.1 Experimental Settings

Datasets. Three real-world datasets are used in our experiments: Diginetica[1], Yoochoose[2], and Nowplaying[3]. Following previous work [2, 8], we pre-processed the three datasets. Specifically, sessions with fewer than 2 items and items appearing less than 5 times were removed. The last week's sessions are used as test data, while the remaining historical data was utilized for training. The dataset is extended and labeled by sequence segmentation, generating multiple session sequences with corresponding labels from session $S = [v_{s,1}, v_{s,2}, \ldots, v_{s,n}]$: $([v_{s,1}], v_{s,2}), ([v_{s,1}, v_{s,2}], v_{s,3}), \ldots, ([v_{s,1}, v_{s,2}, \ldots, v_{s,n-1}], v_{s,n})$. Following [3, 5], We utilize the most recent 1/64 fraction of the training sequences from the Yoochoose dataset. The statistics of the datasets are summarized in Table 1.

Evaluation Metrics and Baselines. We adopt two widely used rank-based metrics: P@K (Precision) and MRR@K (Mean Reciprocal Rank), following previous work [5, 8]. We compare classical traditional algorithms and RNN-based models while also contrasting with state-of-the-art graph neural network models. (i) Traditional models: Item-KNN [1], FPMC [13]. (ii) RNN-based models: GRU4Rec [2], NARM [3], STAMP [5]. (iii) GNN-based models: SR-GNN [8], FGNN [19], DHCN [10].

Parameter Setup. For general settings, we set the batch size to 512, the embedding size to 100, and L_2 regularization to 10^{-5}. For EGNN-SDI, the initial learning rate is set to 0.001. The number of network layers varies between datasets. The baseline model adopts the optimal parameter settings as reported in the original paper.

Table 1. Statistics of the three datasets.

Datasets	Clicks num	Training num	Test num	Items num	Avg length
Diginetica	982961	719470	60858	43907	5.12
Yoochoose 1/64	557248	369859	55989	16766	6.16
Nowplaying	1367963	825304	89824	60417	7.42

5.2 Overall Comparison (RQ1)

Table 2 shows the performance comparison between EGNN-SDI and the baseline algorithms. The top-performing and second-best results are highlighted in bold and underlined, respectively.

The performance of the traditional methods is not ideal. This is because they do not effectively model the item transition. GRU4Rec performs better than FPMC, but is worse

---

[1] https://competitions.codalab.org/competitions/11161.

[2] https://www.kaggle.com/datasets/chadgostopp/recsys-challenge-2015.

[3] http://dbis-nowplaying.uibk.ac.at/#nowplaying.

Table 2. Effectiveness comparison on three datasets.

Model	Diginetica		Yoochoose 1/64		Nowplaying	
	P@20	MRR@20	P@20	MRR@20	P@20	MRR@20
Item-KNN	35.75	11.57	51.60	21.81	15.94	4.91
FPMC	26.53	6.95	45.62	15.10	7.36	2.82
GRU4Rec	29.45	8.33	60.64	22.89	7.92	4.48
NARM	49.28	16.52	68.98	29.61	18.45	6.93
STAMP	45.64	14.32	68.74	29.67	17.66	6.88
SR-GNN	50.73	17.59	70.57	30.94	18.87	7.47
FGNN	50.58	16.84	70.91	30.68	18.78	7.15
DHCN	53.12	18.27	72.14	31.02	22.06	7.90
EGNN-SDI	**53.21**	**18.46**	**73.16**	**31.48**	**22.54**	**8.41**

than Item-KNN on the Diginetica and Nowplaying datasets. This discrepancy emerges because RNNs are optimized for sequence modeling, whereas SBRs present heightened complexity due to potential fluctuations in user preferences throughout sessions.

The subsequent NARM and STAMP models introduced attention mechanisms that significantly improved performance over GRU4Rec. SR-GNN and FGNN outperform RNN-based models, and these improvements can be attributed to the powerful capabilities of graph neural networks. In addition, DHCN incorporates self-supervised learning into network training, resulting in significant improvements in model performance.

Table 3. Impacts of static and dynamic information

Model	Diginetica		Yoochoose 1/64		Nowplaying	
	P@20	MRR@20	P@20	MRR@20	P@20	MRR@20
w/o global	49.22	16.08	68.85	28.30	14.81	5.59
w/o session	52.57	18.14	72.73	31.33	22.38	8.37
w/o SIL	51.21	17.42	72.77	**31.59**	21.92	8.35
w/o DIL	51.21	17.23	70.70	29.77	18.57	6.98
EGNN-SDI	**53.21**	**18.46**	**73.16**	31.48	**22.54**	**8.41**

EGNN-SDI overwhelmingly outperforms all the baselines on three datasets. Unlike other GNN-based models, which utilize only partial session information. For example, SR-GNN only utilizes information from the current session. DHCN although exploits global static information but does not model item transitions in local session. Our approach considers both static and dynamic information, while also integrating relative position information, resulting in superior performance.

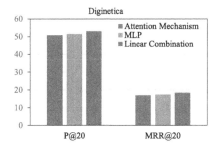 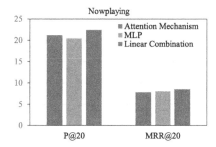

Fig. 4. Performance of different aggregation operations.

5.3 Ablation Study (RQ2)

Next, we conducted ablation experiments to evaluate the effectiveness of static and dynamic information. Specifically, we designed four comparative models. (i) w/o global: EGNN-SDI without static information at global level. (ii) w/o session: EGNN-SDI without static information at session level. (iii) w/o SIL: EGNN-SDI without static information. (iv) w/o DIL: EGNN-SDI without dynamic information.

The results in Table 3 show that the absence of some information worsens the model performance, except for the MRR@20 on the Yoochoose1/64 dataset. This could be attributed to the relatively small size of the Yoochoose 1/64 dataset, which increases the model's susceptibility to overfitting. These results suggest that each type of information learned by the model is indispensable.

5.4 Impact of Aggregation Operations (RQ3)

Since we learn static and dynamic information separately, it is important to evaluate the influence of different aggregation operations (i.e., the attention mechanism, MLP, and linear combination) on EGNN-SDI.

In the case of the attention mechanism, attention scores are computed, normalized, and subsequently used for weighted fusion:

$$\mathbf{H}_f = \text{softmax}(\mathbf{\Theta}(s)\mathbf{w}_{att}) \odot \mathbf{H}_d + \text{softmax}(\mathbf{H}_d\mathbf{w}_{att}) \odot \mathbf{\Theta}(s), \qquad (11)$$

For MLP, only a simple MLP layer is adopted to fuse the two representations of the items, defined as follows:

$$\mathbf{H}_f = \text{MLP}([\mathbf{\Theta}(s) \,\|\, \mathbf{H}_d]), \qquad (12)$$

For the linear combination, we use two learnable parameters α, β to aggregate:

$$\mathbf{H}_f = \alpha \cdot \mathbf{\Theta}(s) + \beta \cdot \mathbf{H}_d, \qquad (13)$$

Figure 4 shows the performance of the different aggregation operations. It's noticeable that despite having additional parameters, both the attention mechanism and the MLP perform worse than linear fusion.

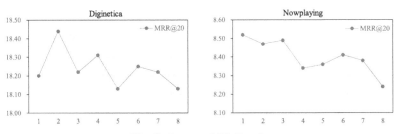

Fig. 5. Impact of SIL Depth.

5.5 Impact of SIL Depth (RQ4)

To study the impact of SIL depth in EGNN-SDI, we vary the numbers of layers of the network within {1, 2, 3, 4, 5, 6, 7, 8}. According to the results presented in Fig. 5, EGNN-SDI is not very sensitive to the number of layers on Diginetica and a two-layer setting is the best. However, on Nowplaying, a single-layer network gives the best performance. Furthermore, the performance on MRR@20 decreases as the number of layers increases. This could be due to the increasingly over-smoothed representations of items.

6 Conclusion

In this paper, we introduce a new model called EGNN-SDI for SBR. Specifically, we propose a new graph neural network enhanced with static and dynamic information. Global and local static information aids the model in understanding global preferences and session context, while dynamic information enables the model to adapt to the user's evolving interests. We design a new item encoder SDI to capture session information from two different perspectives and linearly combine the two representations to improve the feature representation of items. Extensive experiments on three datasets demonstrate that EGNN-SDI surpasses existing methods. Furthermore, ablation studies validate the effectiveness of each component of our model.

In the future, we will try to apply EGNN-SDI to real session-based recommendation scenarios and tune the model.

References

1. Sarwar, B., Karypis, G., Konstan, J., Riedl, J.: Item-based collaborative filtering recommendation algorithms. In: WWW, pp. 285–295 (2001)
2. Hidasi, B., Karatzoglou, A., Baltrunas, L., Tikk, D.: Session-based recommendations with recurrent neural networks. In: ICLR (2015)
3. Li, J., Ren, P., Chen, Z., et al.: Neural attentive session-based recommendation. In: CIKM, pp. 1419–1428 (2017)
4. Tan, Y.K., Xu, X., Liu, Y.: Improved recurrent neural networks for session-based recommendations. In: RecSys, pp. 17–22 (2016)
5. Liu, Q., Zeng, Y., Mokhosi, R., Zhang, H.: STAMP: short-term attention/memory priority model for session-based recommendation. In: SIGKDD, pp. 1831–1839 (2018)

6. Rendle, S., Freudenthaler, C., Schmidt-Thieme, L.: Factorizing personalized Markov chains for next-basket recommendation. In: WWW, pp. 811–820 (2010)
7. Wang, Z., Wei, W., Cong, G., et al.: Global context enhanced graph neural networks for session-based recommendation. In: SIGIR, pp. 169–178 (2020)
8. Wu, S., Tang, Y., Zhu, Y., et al.: Session-based recommendation with graph neural networks. In: AAAI, pp. 346–353 (2019)
9. Xia, X., Yin, H., Yu, J., et al.: Self-supervised graph co-training for session-based recommendation. In: CIKM, pp. 2180–2190 (2021)
10. Xia, X., Yin, H., Yu, J., et al.: Self-supervised hypergraph convolutional networks for session-based recommendation. In: AAAI, pp. 4503–4511 (2021)
11. Ludewig, M., Jannach, D.: Evaluation of session-based recommendation algorithms. User Model. User-Adap. Interact. **28**(4–5), 331–390 (2018)
12. Hu, H., He, X., Gao, J., et al.: Modeling personalized item frequency information for next-basket recommendation. In: SIGIR, pp. 1071–1080 (2020)
13. Eirinaki, M., Vazirgiannis, M., Kapogiannis, D.: Web path recommendations based on page ranking and Markov models. In: CIKM, pp. 2–9 (2005)
14. Xu, K., Li, C., Tian, Y., et al.: Representation learning on graphs with jumping knowledge networks. In: ICML, pp. 5453–5462 (2018)
15. Li, A., Cheng, Z., Liu, F., et al.: Disentangled graph neural networks for session-based recommendation. IEEE Trans. Knowl. Data Eng. **35**(8), 7870–7882 (2023)
16. Vaswani, A., Shazeer, N., Parmar, N., et al.: Attention is all you need. In: Advances in Neural Information Processing Systems, vol. 30 (2017)
17. Pan, Z., Cai, F., Ling, Y., et al.: Rethinking item importance in session-based recommendation. In: SIGIR, pp. 1837–1840 (2020)
18. Yuan, J., Song, Z., Sun, M., et al.: Dual sparse attention network for session-based recommendation. In: AAAI, pp. 4635–4643 (2021)
19. Qiu, R., Li, J., Huang, Z., Yin, H.: Rethinking the item order in session-based recommendation with graph neural networks. In: CIKM, pp. 579–588 (2019)
20. Chang, J., Gao, C., Zheng, Y., et al.: Sequential recommendation with graph neural networks. In: SIGIR, pp. 378–387 (2021)
21. Zhao, S., Wei, W., Zou, D., Mao, X.: Multi-view intent disentangle graph networks for bundle recommendation. In: AAAI, pp. 4379–4387 (2022)
22. Zhang, J., Ma, C., Mu, X., et al.: Recurrent convolutional neural network for session-based recommendation. Neurocomputing **437**, 157–167 (2021)
23. Pang, Y., Wu, L., Shen, Q., et al.: Heterogeneous global graph neural networks for personalized session-based recommendation. In: WSDM, pp. 775–783 (2022)

Construction of Academic Innovation Chain Based on Multi-level Clustering of Field Literature

Cheng Wei$^{(\boxtimes)}$ and Cong Tianshi

Nanjing Agricultural University, Nanjing 210095, China
chengwei@stu.njau.edu.cn

Abstract. Depth exploration and display of the potential correlation of innovation point can be helpful for relevant work such as field innovation discovery and field literature innovation evaluation. First, on the basis of the concept of academic innovation chain, the construction method of academic innovation chain based on multi-level clustering is proposed. Second, combining the text feature mining algorithms of tf-idf, LDA, doc2vec and the Kmeans text clustering algorithm, 639 literatures in the field of "knowledge element" are taken as examples clustering from the three levels of word frequency, topic and semantic. Final, fusion rule method with ALBERT pre-training model to extract the innovation points of literature, then the construction of academic innovation chain is realized. The academic innovation chain connects the originally isolated innovation point linearly. It provides certain references for the research of innovation evaluation and innovation metrics.

Keywords: Academic Innovation Chain · Text Clustering · Innovation Points Extraction · Innovation Evaluation

1 Introduction

Innovation stands as an indispensable requirement in scientific research, with the extraction, measurement, and evaluation of innovative knowledge emerging as prominent research areas within academic evaluation [19,22]. It encompasses the discovery or creation of novel knowledge within the existing knowledge framework, characterized by both relevance and fundamental divergence from pre-existing knowledge [5,6].

Various methods for analyzing literature associations, such as keyword co-occurrence networks [3,7], citation networks [12], and topic evolution networks [20], have been widely employed and yielded promising results. However, these methods encounter challenges in visualizing potential connections between innovation points [7,12,20], thereby limiting the comprehensive assessment of innovative contributions.

The clustering algorithm, a versatile and powerful tool, is widely applied across diverse domains for data analysis and pattern recognition. Notably, Pen

Z. Wang and C. W. Tan (Eds.): PAKDD 2024 Workshops, LNAI 14658, pp. 82–94, 2024.
https://doi.org/10.1007/978-981-97-2650-9_7

et al. employ the Self-Organizing Map clustering algorithm for image segmentation [16]. Additionally, clustering algorithms are invaluable for anomaly detection
[14,21]. Moreover, clustering methods play a pivotal role in social network analysis, as elucidated by Curiskis et al. [4]. These collective efforts underscore the
efficacy of clustering algorithms across various applications.

Text clustering, employing various clustering algorithms for analyzing text
data, especially within the domains of topic clustering [4,15] and semantic clustering [8,11], represents an effective strategy for categorizing similar texts. This
approach provides a tangible means of organizing innovative concepts dispersed
throughout diverse literature sources. As such, the principal aim of this study is
to consolidate interconnected literature using text clustering algorithms. Additionally, our objective extends to establishing a chronological framework for innovation points by systematically extracting them from the literature in accordance
with their order of publication.

This study introduces an innovative method for constructing academic innovation chains through multi-level clustering. We focus on Chinese journal literature in the knowledge element field as our experimental domain. Leveraging
tf-idf, LDA, and doc2vec algorithms, we achieve feature representation of literature across three levels: word frequency, topic, and semantic. Subsequently, text
clustering is performed using the kmeans algorithm. To enhance the process, we
integrate a fusion rule method with the ALBERT pre-training model to extract
innovation points from the literature. Finally, we not only construct but also
visualize academic innovation chains.

The primary contribution of this study lies in its ability to observe innovation
points within a progressive and dynamic chain structure. This innovative approach holds significant implications for informing innovation assessment, knowledge organization, and related studies.

2   Methodology

2.1   Overview of Methodology

This study proposes an academic innovation chain construction method based on
multi-level clustering of field literature is shown in Fig. 1, including the following
steps.

1. Determine the scope of the field and obtain the field literature from academic
   databases. Taking the titles and abstracts of the literature as data objects,
   the title and abstract corpus is constructed by data preprocessing through
   separate words, delete stopwords and lexical labeling using the jieba library.
2. Utilizing the title and abstract corpus, the textual features of the field literature are mined in-depth from three levels, namely, word frequency, topic and
   semantic, based on the tf-idf, LDA and doc2vec algorithms, respectively. In
   turn, the feature vector representation of the literature is formed at different
   levels.

3. Based on the Kmeans text clustering algorithm, the feature vector representations of the field literature are clustered at three levels, namely, word frequency, topic and semantics, respectively. Clustering clusters of field literature at three different levels are obtained.
4. Extract the innovation points in the abstracts of field literature by rule and manually, and construct a categorized corpus composed of innovation points and non-innovation points. Learning textual features of innovation points in the field based on lightweight pre-trained model ALBERT through sentence-level binary classification method to realize automatic extraction of innovation points in the field literature.
5. Taking the innovation points as the basic unit, the innovation points of the literature of the same type of clusters are chained and organized according to the chronological order of the publication of the literature, so as to realize the construction of the academic innovation chain.

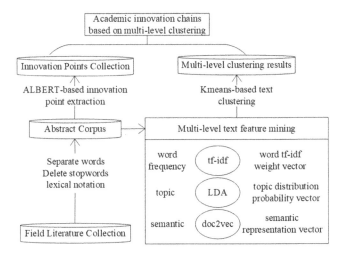

Fig. 1. Academic innovation chain construction method

2.2   Multi-level Text Feature Mining Algorithms

Word Frequency. Tf-idf algorithm does not pay attention to the order and dependency of the words between the word, only from the point of view of the frequency of the word to determine a word's probability of occurrence. For any word in any document, its tf-idf weight is the product of tf (the number of occurrences of the word in the current document) and idf (the total number of documents divided by the number of words containing the word, the result is taken as a logarithm). Assuming that the collection contains N documents with a total of M non-repeating words, a N*M matrix can be constructed based on the tf-idf algorithm, which is a highly sparse matrix.

Topic. LDA (Latent Dirichlet Allocation) topic model is commonly used to infer the topic distribution of documents. LDA first needs to confirm the value of the number of topics k. The best k value is usually determined by the confusion value. Then the probability of distribution of words in k topics is obtained through the computation and training of the corpus. Finally, the topic distribution probabilities are calculated from the words contained in the documents. A k-dimensional document-topic distribution vector is constructed to characterize the topic content of the document, and the elements of the vector represent the probability that the document belongs to each topic [1,9].

Semantic. Doc2vec integrates the dependency of a word with its preceding and following words as well as the order of occurrence of words. In the training process, a certain part of the document is intercepted as a prediction word, and the word vectors of other words are used as input. And in each round of training, the sentence vector D of the current document is taken as the feature vector input, and D is characterized as the semantic content of the document based on the inter-word semantic dependency. D is continuously updated during iterative training so that the model learns the missing content in the current context as well as the main idea content of the paragraph. Finally, a semantic vector W is generated for each word in the document, and a semantic vector D is generated for each document [10].

2.3   Kmeans Clustering Algorithm

Kmeans is one of the most commonly used text clustering algorithms. The processing steps are as follows: first, based on the sample set S, randomly initialize k clustering centroids, which can be randomly selected from the sample set S or randomly generated. Next, traverse each sample x of the sample set S, calculate the space vector distance from x to the k centroids, and assign x to the class cluster y with the shortest distance. Then, the vector mean of all samples in each class cluster is calculated as the new clustering centroid. Finally, the above two steps are repeated until a certain number of iterations is reached or the clustering centroids are no longer changed [13,18].

2.4   Innovation Points Extraction

Innovation points extraction is a sentence-level text categorization task with the following process:

1. Segment the abstract text into sentences.
2. For the clauses after clause splitting, the preliminary extraction of innovation points in the abstract is realized based on the trigger word rule. Some of the rules and examples are shown in Table 1.
3. Manually check and correct the results of the rule-based innovation extraction, and construct a "literature innovation points" combination and classification corpus.

4. Adopt TensorFlow framework, create neural network model using Keras, and construct automatic extraction model of field innovation points based on lightweight pre-training model ALBERT to realize the generalized extraction of innovation points in field literature.

Table 1. Partial rules and examples of innovation points extraction.

Trigger word	Example of extraction results
propose	An oil spill incident scenario model based on key scenario driver elements is proposed.
explore	The basic methods of structured processing of geological data texts are explored.
construct	Constructing an ontological knowledge base of chest paralysis evidence.
create	Creating a structured body of knowledge linked by concepts.
improve	Improving Bootstrapping methods

3   Empirical Research

3.1   Data Collection and Preprocessing

In this study, we take the field of "knowledge element" as an example to explore the feasibility of the academic innovation chain construction method in the proposed field. And 639 field literatures are retained, spanning from 1981 to 2022.

The determination of k value in Kmeans algorithm is very important work. The k-value interval is set to [2,50], and the values are traversed for multiple rounds of clustering, and the best k-value is selected based on the following two ways. (i) Silhouette Coefficient is used to evaluate the clustering effect and as a reference for selecting the optimal k-value [17]. (ii) Tthe T-SNE visualization dimensionality reduction algorithm is used to map the multi-dimensional text feature vectors into a two-dimensional space. The reasonableness of the clustering results is evaluated manually to select the best k value.

3.2   Multi-level Clustering Experiments

Word Frequency. There are 5341 non-repeated words in the corpus. A 2D matrix of 639*5341 is obtained. In turn, Kmeans clustering is implemented based on sklearn library. The largest Silhouette Coefficient is 0.0207 (k takes the value of 50), and the smallest Silhouette Coefficient is 0.007 (k takes the value of 2), and the Silhouette Coefficient tends to 0. This indicates that there is a strong knowledge linkage within the field of knowledge elements, and when the dataset is classified into multiple class clusters, there is a certain amount of variability in the sample data between different class clusters compared to that in the same class clusters, but it is not particularly prominent, which objectively results in the effect of taking k value based on Silhouette Coefficient is not ideal. Therefore, it is manually judged that the overall clustering effect reaches the best when k is 4. The visualization of tf-idf+Kmeans clustering results based on T-SNE dimensionality reduction is shown in Fig. 2. Number the individual clusters as T0, T1, T2, T3.

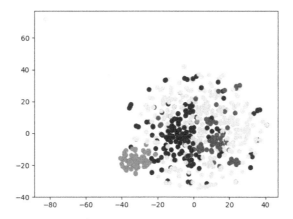

Fig. 2. Visualization of tf-idf+Kmeans clustering results based on T-SNE dimensionality reduction.

In Fig. 2, the T3 category (yellow nodes) has the largest number of samples, which causes some interference to the correct categorization of the samples of the other clusters, and there is a problem of unclear boundaries with other clusters as well as mutual confusion. Therefore, its clustering results have some limitations. In contrast, the other three class clusters have clearer boundaries with each other.

Topic. The LDA topic model is invoked using the gensim library. The number of passes through the corpus during training is 10, and the prior alpha of the document-topic distribution and the prior of the topic-word distribution are both set to auto. After manual judgment, the integers in the interval [5,25] are used as the candidate topic numbers for iterative experiments. Perplexity is calculated in each round of experiments, and the inflection point of perplexity is used to set the optimal number of topics, and the topic model achieves the best effect when the number of topics is 15. The corpus is transformed into a 2D matrix of 639*15. During the iterative clustering process, the difference in Silhouette Coefficient in each round of experiments is small, and when k is 14, the overall clustering effect is judged manually to be optimal. The visualization of LDA+Kmeans clustering results based on T-SNE dimensionality reduction is shown in Fig. 3. Number the individual clusters as L0, L1, ..., L12, L13.

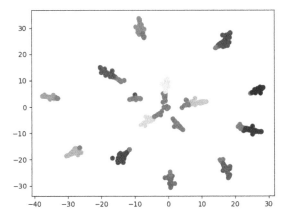

Fig. 3. Visualization of LDA+Kmeans clustering results based on T-SNE dimensionality reduction.

In Fig. 3, the 15 topics modeled based on LDA form 15 "clusters" with obvious differentiation in the spatial vector distribution, except for L7 (dark green nodes), which has the largest number of samples and its samples are distributed in multiple "communities". The samples in the other clusters are distributed in relatively fixed "communities", and the differences in topic characteristics are more obvious.

Semantic. The doc2vec sentence vector semantic training model is invoked using the gensim library. The sentence vector dimension is defined as 100 dimensions, and the window value is set to 3. The minimum word frequency is set to 1, and the number of iterations is set to 100. A word vector representation of 5,341 words is obtained, as well as a two-dimensional matrix of 639*100. In the iterative clustering process, the difference between the Silhouette Coefficient in each round of experiments is still small, and when k is 7, the overall clustering effect is judged manually to be optimal. The visualization of doc2vec+Kmeans clustering results based on T-SNE dimensionality reduction is shown in Fig. 4. Number the individual clusters as D0, D1, ..., D5, D6.

In Fig. 4, the semantic-based clustering results are relatively the most obvious differentiation, and the main content of the literature within each category cluster is more unified. It can also be found that the distribution of samples of the same category is relatively centralized and clear, except for a certain amount of confusing interference at the boundaries of each cluster.

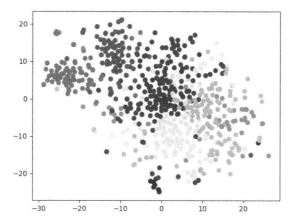

Fig. 4. Visualization of doc2vec+Kmeans clustering results based on T-SNE dimensionality reduction.

3.3 Innovation Points Extraction Experiments

The innovation points classification corpus contains 1982 positive samples for innovation points clauses and 4288 negative samples for non-innovation points clauses. The training set and test set are divided according to 9:1. The pre-training model is Google's open-source Albert large Chinese pre-training model, Albert-large-zh. The training parameters are adjusted as follows. The batch size is 128; the maximum text length is 128; the epoch is 10; the learning-rate is 0.00001. The performance of the innovation points auto-extraction model on the test set is evaluated using the performance measures commonly used in text categorization: precision (P), recall (R), and the F1-score (F1), and the results of the calculations are retained in two decimal places. The evaluation results are shown in Table 2.

Table 2. The evaluation results of automatically innovation points extracted models.

Classification	Micro-P	Micro-R	Micro-F
innovation points	95.73%	79.29%	86.74%
non-innovation points	91.13%	98.36%	94.61%

The overall classification accuracy of the auto-extraction model on the test set with 198 positive samples and 428 negative samples is 92.33%, which achieves good results. The accuracy of automatic extraction of innovation points is higher, while the recall is not yet 80%. It shows that the extracted innovation points have high confidence, but it leads to about one-fifth omission. Although there are certain defects, the model can assist the automatic extraction of field innovation points.

3.4   Constraction of Academic Innovation Chain

Comparing the clusters to which different innovation points belong at different levels, there are 583 literatures with the same clusters as at least one literature whose innovation points belong to the same clusters at the three levels of word frequency, topic and semantics. They can be considered to be highly correlated with each other with unified multi-level features, and the combination of the three clusters at the three levels is called the public cluster combination. There are 119 such public cluster combinations such as "T3-L7-D3".

Taking the public cluster combination "T3-L2-D4" as an example, its academic innovation chain is visualized in Fig. 5, in which the innovation points are basically the proposed special object-oriented knowledge element representation, extraction and association methods. It reflects the high correlation between a series of innovation points and also provides a visualization method for the comparative assessment of the differences of innovation points.

Proposing a cross-lingual knowledge unit transfer method (SHACUT) based on semantic hierarchical modeling.

Integrating Bayesian conditional probability with semantic web, proposing a novel knowledge representation method.

Constructing a knowledge model for emergency decision-making based on knowledge elements.

Constructing a tree-based knowledge map model that encompasses three fundamental elements.

Proposed a mathematical model for knowledge metadata fusion and integration of heterogeneous metadata.

A design knowledge repository system based on ontology can discover relevant design knowledge.

Extracting product design methods by expressing conceptual information through structured knowledge units.

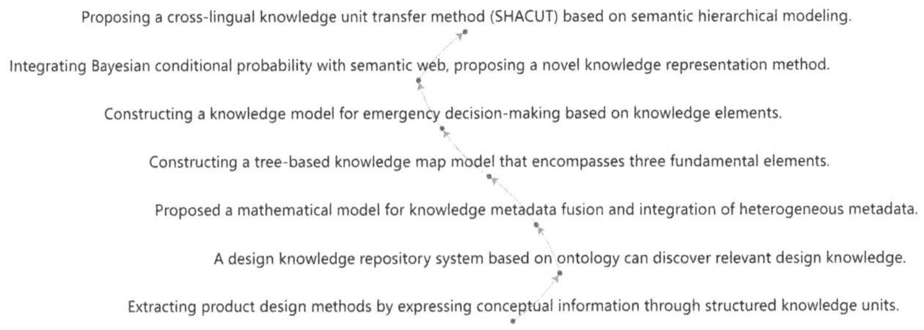

Fig. 5. Academic innovation chain based on "T3-L2-D4" public cluster combination.

Taking the innovation points of the literature "Research on the Extraction Rules of Knowledge Element Contents in the South China Sea Documents of the Republic of China Period" as an example, its localized academic innovation chain examples on three feature levels are shown in Fig. 6, where the green connecting line indicates the correlation based on the word frequency feature, the red connecting line indicates the correlation based on the topic feature, and the blue connecting line indicates the correlation based on the semantic feature. In Fig. 6, it shows other related innovation points. Such as the knowledge element entity and relationship extraction method based on word frequency features, the knowledge network fusion method based on topic features, and the knowledge graph construction model and integrated construction method based on semantic features. It reveals the potential correlation between the studies at different levels, and also intuitively reflects the essential differences between the related studies, highlight the different innovation points.

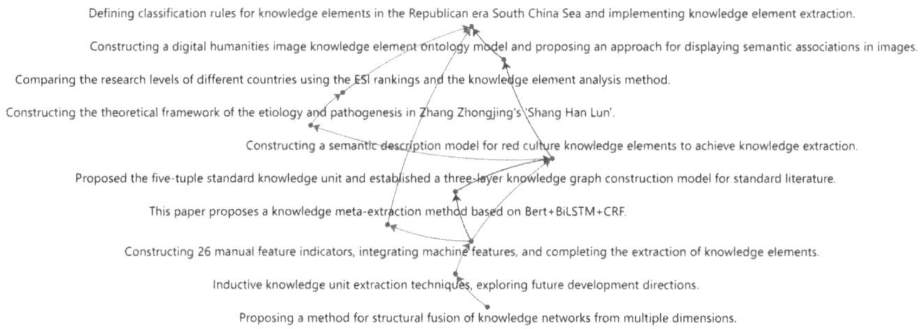

Defining classification rules for knowledge elements in the Republican era South China Sea and implementing knowledge element extraction.

Constructing a digital humanities image knowledge element ontology model and proposing an approach for displaying semantic associations in images.

Comparing the research levels of different countries using the ESI rankings and the knowledge element analysis method.

Constructing the theoretical framework of the etiology and pathogenesis in Zhang Zhongjing's 'Shang Han Lun'.

Constructing a semantic description model for red culture knowledge elements to achieve knowledge extraction.

Proposed the five-tuple standard knowledge unit and established a three-layer knowledge graph construction model for standard literature.

This paper proposes a knowledge meta-extraction method based on Bert+BiLSTM+CRF.

Constructing 26 manual feature indicators, integrating machine features, and completing the extraction of knowledge elements.

Inductive knowledge unit extraction techniques, exploring future development directions.

Proposing a method for structural fusion of knowledge networks from multiple dimensions.

Fig. 6. Example of academic innovation chain based on multi-level clustering.

3.5   Comparison Experiment

References provide reference and borrowing for new literature, and promote or facilitate the generation of new literature [2, 12]. Therefore, citation networks are often used for literature association analysis. As a comparative experiment, the citation networks of 639 literature are constructed as shown in Fig. 7.

In Fig. 7, the citation network can be used to quickly discover high-impact papers in the field, as well as to discover the linkages between the literature. It is able to better show the literature associations globally. Citation networks are also relatively easy to build. However, it is difficult to visualize the linkages between the literature at the content level, especially at the innovation points. And it is difficult to reflect the degree of differences in literature associations. Therefore, academic innovation chain can be used as a complement to citation networks to assist in field innovation knowledge discovery and assessment.

4   Discussion

Academic innovation chains linearly correlate otherwise isolated innovation points. It can either organize highly relevant innovation points in an orderly manner based on the potential correlation of the unity of textual features at multiple levels, or organize relevant innovation points from the perspective of the correlation of textual features at different levels. The academic innovation chain clearly and intuitively shows the evolution path and characteristic trend of innovation in chronological order, providing a visualization tool for the comparative analysis of the relevance and difference between innovation points in the field. It can help scholars to grasp the historical research trend and inspire their innovation selection and discovery, and it can also assist experts to carry out innovation evaluation more efficiently and objectively. In addition, when a batch of new field literature are published, the academic innovation chain based on multi-level clustering can be quickly updated. It can not only detect the most cutting-edge innovations in the field in time, but also assess the innovation value based on the force of innovation points on the updating of the academic innovation chain.

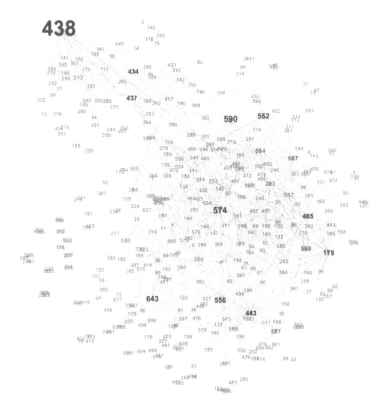

Fig. 7. Literature association network based on citation relationships.

5   Conclusion and Future Work

This study explores the academic innovation chain construction method based on multi-level clustering of field literature. Combining the text feature mining algorithms of tf-idf, LDA, doc2vec and the Kmeans text clustering algorithm, the field literature is clustered from the three levels of word frequency, topic and semantics, respectively. Then the academic innovation chain under the time dimension is constructed on the basis of innovation points extraction by rule method and ALBERT pre-training model. The feasibility of the method is verified by taking the knowledge element field as an example. The potential value of academic innovation chain for innovation discovery and innovation assessment is explored, which provides a new perspective for related research.

This study also has the following limitations. More algorithms are not introduced in the clustering process for comparative experiments to achieve better clustering results. In the clustering process, only the title and abstract text are used, and the full text of literature, which contains richer semantic information, is not used. This will be the direction of improvement for subsequent research. Further, in the subsequent experiments we will use larger scale data for more comprehensive innovative knowledge mining and organization to build a practical experimental platform. We will also optimize the construction method and display form of the academic innovation chain through quantitative evaluation and qualitative evaluation.

References

1. Blei, D.M., Ng, A.Y., Jordan, M.I.: Latent dirichlet allocation. J. Mach. Learn. Res. **3**, 993–1022 (2003)
2. Boyack, K.W., Klavans, R.: Co-citation analysis, bibliographic coupling, and direct citation: which citation approach represents the research front most accurately? J. Am. Soc. Inform. Sci. Technol. **61**, 2389–2404 (2010)
3. Cui, J., Wang, Z., Ho, S.B., Cambria, E.: Survey on sentiment analysis: evolution of research methods and topics. Artif. Intell. Rev. **56**, 8469–8510 (2023)
4. Curiskis, S.A., Drake, B., Osborn, T.R., Kennedy, P.J.: An evaluation of document clustering and topic modelling in two online social networks: Twitter and reddit. Inf. Process. Manage. **57**, 102034 (2020)
5. Ghosal, T., Edithal, V., Ekbal, A., Bhattacharyya, P., Chivukula, S.S.S.K., Tsatsaronis, G.: Is your document novel? Let attention guide you. an attention-based model for document-level novelty detection. Inf. Process. Manage. **27**, 427–454 (2021)
6. Ghosal, T., Saikh, T., Biswas, T., Ekbal, A., Bhattacharyya, P.: Novelty detection: a perspective from natural language processing. Comput. Linguist. **48**, 77–117 (2022)
7. Grames, E.M., Stillman, A.N., Tingley, M.W., Elphick, C.S.: An automated approach to identifying search terms for systematic reviews using keyword co-occurrence networks. Methods Ecol. Evol. **10**, 1645–1654 (2022)
8. Huang, J., Gong, S., Zhu, X.: Deep semantic clustering by partition confidence maximisation (2020)
9. Jelodar, H., et al.: Latent Dirichlet allocation (LDA) and topic modeling: models, applications, a survey. Multimedia Tools Appl. **78**, 15169–15211 (2019)
10. Kim, D., Seo, D., Cho, S., Kang, P.: Multi-co-training for document classification using various document representations: TF-IDF, LDA, and Doc2Vec. Inf. Sci. **477**, 15–29 (2019)
11. Kim, S., Park, H., Lee, J.: Word2vec-based latent semantic analysis (W2V-LSA) for topic modeling: a study on blockchain technology trend analysis. Expert Syst. Appl. **152**, 113401 (2020)
12. Kleminski, R., Kazienko, P., Kajdanowicz, T.: Analysis of direct citation, co-citation and bibliographic coupling in scientific topic identification. J. Inf. Sci. **48**, 349–373 (2022)
13. Kodinariya, T.M., Makwana, P.R.: Review on determining number of cluster in K-means clustering. Int. J. **1**, 90–95 (2013)
14. Li, J., Izakian, H., Pedrycz, W., Jamal, I.: Clustering-based anomaly detection in multivariate time series data. Appl. Soft Comput. **100**, 106919 (2021)
15. Onan, A.: Two-stage topic extraction model for bibliometric data analysis based on word embeddings and clustering. IEEE Access **7**, 145614–145633 (2019)
16. Pen, H., Wang, Q., Wang, Z.: Boundary precedence image inpainting method based on self-organizing maps. Knowl.-Based Syst. **216**, 106722 (2021)

17. Shahapure, K.R., Nicholas, C.: Cluster quality analysis using silhouette score. In: 2020 IEEE 7th International Conference on Data Science and Advanced Analytics (DSAA), pp. 747–748. IEEE (2020)
18. Steinley, D.: K-means clustering: a half-century synthesis. Br. J. Math. Stat. Psychol. **59**, 1–34 (2006)
19. Wang, J., Ma, X., Zhao, Y., Zhao, J., Heydari, M.: Impact of scientific and technological innovation policies on innovation efficiency of high-technology industrial parks-a dual analysis with linear regression and QCA. Int. J. Innov. Stud. **6**, 169–182 (2022)
20. Wang, X., Wang, H., Huang, H.: Evolutionary exploration and comparative analysis of the research topic networks in information disciplines. Scientometrics **126**, 4991–5017 (2021)
21. Wang, Z., Tong, V.J.C., Xin, X., Chin, H.C.: Anomaly detection through enhanced sentiment analysis on social media data. In: 2014 IEEE 6th International Conference on Cloud Computing Technology and Science, pp. 917–922. IEEE (2014)
22. Wei, X., Shen, L.: A research review of the academic paper innovativeness. Documentat., Inf. Knowl. **39**, 68–79 (2022)

DLVS4Audio2Sheet: Deep Learning-Based Vocal Separation for Audio into Music Sheet Conversion

Nicole Teo[1], Zhaoxia Wang[1]([⊠]), Ezekiel Ghe[1], Yee Sen Tan[1],
Kevan Oktavio[1], Alexander Vincent Lewi[1], Allyne Zhang[1],
and Seng-Beng Ho[2]([⊠])

[1] Singapore Management University, 80 Stamford Rd, Singapore 178902, Singapore
{nicolet.2023,zxwang,ezekiel.ghe.2020,yeesen.tan.2020,
kevano.2020,avlewi.2021,allynezhang.2021}@smu.edu.sg
[2] Institute of High Performance Computing, A*STAR, 1 Fusionopolis Way,
Singapore 138632, Singapore
hosb@ihpc.a-star.edu.sg

Abstract. While manual transcription tools exist, music enthusiasts, including amateur singers, still encounter challenges when transcribing performances into sheet music. This paper addresses the complex task of translating music audio into music sheets, particularly challenging in the intricate field of choral arrangements where multiple voices intertwine. We propose DLVS4Audio2Sheet, a novel method leveraging advanced deep learning models, Open-Unmix and Band-Split Recurrent Neural Networks (BSRNN), for vocal separation. DLVS4Audio2Sheet segments choral audio into individual vocal sections and selects the optimal model for further processing, aiming towards audio into music sheet conversion. We evaluate DLVS4Audio2Sheet's performance using these deep learning algorithms and assess its effectiveness in producing isolated vocals suitable for notated scoring music conversion. By ensuring superior vocal separation quality through model selection, DLVS4Audio2Sheet enhances audio into music sheet conversion. This research contributes to the advancement of music technology by thoroughly exploring state-of-the-art models, methodologies, and techniques for converting choral audio into music sheets. Code and datasets are available at: https://github.com/DevGoliath/DLVS4Audio2Sheet.

Keywords: Music · Choral audio · Music sheet · Vocal separation · Audio-to-Sheet · Deep learning · Open-Unmix · Band-Split Recurrent Neural Networks (BSRNN)

1 Introduction

In the dynamic realm of music technology, the transformation of intricate audio into accurate music sheets is a perpetual and significant endeavor [2]. This task,

Z. Wang and C. W. Tan (Eds.): PAKDD 2024 Workshops, LNAI 14658, pp. 95–107, 2024.
https://doi.org/10.1007/978-981-97-2650-9_8

often termed automatic music transcription, is also a challenging endeavor at the intersection of signal processing and artificial intelligence (AI) [1]. It involves developing computational methods to convert acoustic music data into various forms of musical notation [27,28].

The complexity of automatic music transcription is particularly pronounced in choral music, where multiple voices harmonize to create a rich auditory tapestry [2]. Unlike popular music recordings, which often isolate instruments and vocals, choral compositions inherently involve multiple voices blending together to form intricate harmonie [12,14].

Choral music, known for its polyphonic nature, often involves intricate layering of multiple voices singing in various pitches and timbres [2,14]. This complexity makes it challenging to precisely differentiate between the audio source components such as Soprano, Alto, Tenor, and Bass (SATB) sections [2]. These challenges underscore the need for further exploration and development of techniques tailored to choral music source separation.

Music source separation, particularly in vocal separation, has a rich history predating the emergence of deep learning. Early traditional algorithms, such as Independent Component Analysis (ICA) and Hidden Markov Models (HMMs), established the foundation of this field [10].

The emergence of machine learning, particularly deep learning methodologies, has spurred significant advancements across diverse domains [8,9,13,15, 21,22]. Hu et al. employ deep learning for forecasting stock market trends [8] and sentiment analysis [9]. Tan et al. apply deep learning techniques in video analysis [21], while Teo et al. utilize them for aspect-based sentiment analysis tasks [22]. Additionally, Ni et al. provide a comprehensive survey on deep learning based dialogue systems [13]. These endeavors collectively underscore the versatility and efficacy of deep learning across a spectrum of applications.

In recent years, the field of music research has undergone a paradigm shift, veering away from traditional methodologies to embrace the integration of deep learning models [2,11,27]. However, the adoption of deep learning-based approaches remains somewhat constrained, particularly within the intricate domain of choral arrangements, where the intertwining of multiple voices poses significant challenges.

In this paper, we present DLVS4Audio2Sheet, a novel approach that utilizes deep learning models, including Open-Unmix [20,23] and Band-Split Recurrent Neural Networks (BSRNN) [11], to accurately segment choral music into distinct audio sections. DLVS4Audio2Sheet is specifically designed to overcome challenges associated with converting audio into music sheets.

The following is a summary of our contributions:

1. This research propose a new method, DLVS4Audio2Sheet, which leverages and evaluates two advanced models, Open-Unmix and Band-Split Recurrent Neural Networks (BSRNN), to tackle the complex task of transcribing choral music audio into notated music sheets.
2. This research offers practical insights into the difficulties that arise while managing the complexity of choral arrangements. Creative ways are also suggested

to get beyond these obstacles, offering innovative techniques for professionals and scholars involved in source separation of choral music.

3. This research contributes to a thorough comprehension of the advantages and disadvantages of the evaluated deep learning models in the particular setting of source separation for choral music. This can be an important resource for the larger community of music technology scholars and practitioners.

2   Related Work

Automatic music transcription systems have the potential to revolutionize various interactions between individuals and music, spanning music education, creation, production, search, and musicology [1,17]. This technology serves as a catalyst for societal and economic impacts. Music source separation, which involve estimating and inferring source signals from mixed data, are closely intertwined with automatic music transcription, highlighting its broader relevance and applications [28].

The task of music source separation, such as vocal separation, presents a significant opportunity for enhancing automatic music transcription systems. [19]. By effectively separating individual sources within complex audio recordings, such as vocals, and background noise, these systems can more accurately transcribe the underlying musical notes [12]. This not only improves the fidelity of the transcription process but also opens up new avenues for applications in music education, production, and analysis [11].

The field of music source separation, including vocal separation, has a long history that predates the deep learning era, with traditional techniques such as ICA and HMMs [10]. In addition, music source separation has undergone a revolution thanks to the development of deep learning algorithms [6]. More recently, Transformer models, Recurrent Neural Networks (RNNs), and Convolutional Neural Networks (CNNs) have demonstrated remarkable achievements in accurately identifying sources, allowing for more subtle separation even in complex audio mixes [11,12,23]. Although these methods were groundbreaking at the time, they have trouble managing overlapping and complex audio sources, which is a major obstacle in the separation of choral singing [7].

Tzinis et al. introduced an efficient neural network for end-to-end audio source separation, leveraging the SuDoRMRF backbone structure and one-dimensional convolutions for feature aggregation. Despite its simplicity, the model achieves high-quality separation with minimal computational resources, outperforming state-of-the-art approaches. However, it is worth noting that the method is limited to speech and environmental sound datasets [24]. Chandna demonstrated the potential of deep learning approaches in choir ensemble separation by training and evaluating state-of-the-art source separation algorithms using publicly available choral singing datasets [2].

Notable models such as Open-Unmix and Band-Split RNN show promise in music transcription domains, with Open-Unmix excelling in isolating vocals and Band-Split RNNs adept at discerning similar frequency ranges [11,20].

Despite their advancements, overlapping and complex audio sources, particularly in choral singing, remain problematic [7].

This research aims to leverage advanced models to enhance the effectiveness of music transcription processes, addressing these persistent challenges.

3   Methodology

We develop DLVS4Audio2Sheet, a novel method that leverages two deep learning models, Open-Unmix and BSRNN, to convert choral music audio into notated music sheets. DLVS4Audio2Sheet undergoes evaluation through two distinct modules, allowing for a comprehensive assessment of each individual model's advantages and disadvantages in the context of source separation for choral music. Based on the outcome of this analysis, the most effective models are selected as components of the DLVS4Audio2Sheet method for further processing.

3.1   Overall Design of the DLVS4Audio2Sheet Method

As depicted in Fig. 1, our proposed DLVS4Audio2Sheet method consists of two sub-modules: (A) Vocal Separation and (B) Audio to Music Sheet. In sub-module (A) Vocal Separation, two distinct deep learning models, Open-Unmix and BSRNN, are employed, each tailored for processing choral music inputs. Notably, separate training is conducted for each section of the choir, resulting in distinct trained models for individual sections. Following training, the performance of each model is rigorously tested and evaluated for every choir section. Subsequently, the model with the highest evaluation score is selected for further processing.

It is essential to highlight that the output of the selected deep learning models in the Vocal Separation sub-module (A) results in four distinct audio files corresponding to specific choir sections: Soprano, Alto, Tenor, and Bass. These audio files serve as inputs for the Audio to Music Sheet sub-module (B), where they undergo a conversion procedure to transform them into music score sheets. The audio file with the highest evaluation score is chosen for the next step in the process. This systematic approach ensures the accurate transcription of choral music into notated form, facilitating comprehensive analysis and interpretation.

3.2   Vocal Separation Leveraging the Two Deep Learning Models: Open-Unmix and BSRNN

In the overall methodology design of the DLVS4Audio2Sheet method (Fig. 1), vocal separation is identified as one of the core modules. Two individual deep learning-based models are selected to be components of the DLVS4Audio2Sheet method after undergoing thorough evaluation.

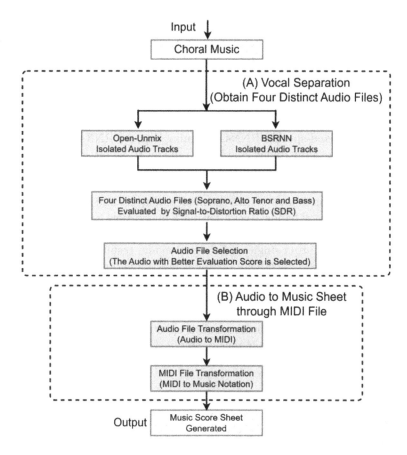

Fig. 1. Overall Design of the DLVS4Audio2Sheet Method

Open-Unmix Model: In the open-unmix model, preprocessing involves feeding choral music as input and cropping the audio signal's frequency range to focus on relevant frequencies. This is followed by Short-Time Fourier Transform (STFT) to convert the signal into the frequency domain, and normalization to stabilize the signal for neural network processing.

Neural network processing begins with spectral feature extraction to identify crucial features from the frequency spectrum. Source estimation then predicts distinct sources in the audio stream, followed by masking to isolate specific audio components.

Postprocessing includes converting the frequency-domain signal back to the time domain using inverse Short-Time Fourier Transform (ISTFT) to reconstruct the audio waveform. Phase reconstruction ensures proper temporal alignment and coherence of separated audio tracks. This process generates isolated audio tracks from the original choral music input [20].

Band-Split RNN (BSRNN): The BSRNN [11] model introduces an innovative approach to music separation through a "band-split" strategy, intricately dividing input audio spectrograms into sub-bands across different frequency levels. This technique leverages the distinct frequency allocations for each SATB choir section to differentiate between them. Unlike other models focusing on variables like timbre and reverberation, BSRNN prioritizes sub-frequencies, making it suitable for our objective.

Choral music is initially subjected to frequency decomposition during pre-processing. The decomposed frequencies then undergo processing via Fully-Connected layer band-specific RNNs, assigning distinct RNNs to different frequency bands. This adjustment enables the model to adapt its processing and learning to the characteristics of various frequency ranges, enhancing audio separation precision.

Feature extraction follows, where the model identifies and extracts significant features from frequency bands crucial for mask generation. Subsequently, masks are generated, and an inverse frequency transformation is applied to combine the processed bands into a single audio signal again, resulting in isolated audio recordings. This process effectively distinguishes the various elements of choral music.

BSRNN employs a dual loss function, combining Mean Absolute Error (MAE) in both the frequency and time domains, ensuring preservation of time-domain signal integrity and capturing subtle spectral details. This design enable it suitable and effective for choir music, emphasizing the intricate interactions between temporal and frequency elements in choir compositions, facilitating clear and accurate vocal separation.

3.3   Audio to Score Sheet Through MIDI File

As depicted in the overall methodology design of the DLVS4Audio2Sheet method (Fig. 1), the Audio to Music Sheet module (Module (B)) is another core component. The output of the Vocal Separation module (Module (A)) serves as the input for the Audio to Music Sheet module (Module (B)).

To convert vocal tracks obtained from deep learning models into musical notes, we utilize Python libraries or existing methods such as Librosa, Aubio, SciPy, and versatile music21 [5]. These methods employ probabilistic modeling techniques to infer the most likely sequence of musical states, resulting in an intermediate piano-roll representation detailing note onsets, offsets, pitches, and names. Subsequently, a MIDI file is generated, incorporating tempo information. In our research, we employ music21 and AnthemScore software to convert segmented audio files into score sheets, simplifying access for musicians and choral groups.

4   Dataset

4.1   Scarcity of Datasets

One obstacle in investigating machine learning methods for multiple music voice separation is the absence of a suitable and large annotated dataset. In order to get around this problem, we created our own multi-track dataset for training by combining different combinations from multi-track datasets that already exist. This is a common data augmentation method used in various multiple singing voice separation papers [4, 18] to address the lack of data. In this section, we will cover the datasets used, and explain the data augmentation process.

4.2   Raw Datasets

Table 1 shows the datasets used in this study. They are all freely accessible through public sources or have been published as parts of other academic works [2, 3].

Table 1. Existing datasets for choral music separation [2,3]

Dataset	No. of songs	Duration (minutes)
Choral Singing Dataset	3 songs	7
ESMUC Choir Dataset	3 songs	31
Cantoria Dataset	11 songs	20

All three datasets comprise full-length choir songs consisting of vocal sections including Soprano, Alto, Tenor, and Bass sections. Each individual performer was recorded using a close-up microphone, while distant microphones simultaneously captured the entire choir's sound. Consequently, the separate audio recordings were processed for each solo singer as well as recordings of the full choir singing in unison ('mixture') for every song. When utilizing the complete choir recording as input, the individual recordings serve as the ground truth or reference for the model's output.

While the ESMUC Choir Dataset was sampled at 22 KHz using mono channels, the Choral Singing Dataset and Cantoria Dataset were sampled at 44 KHz using stereo audio channels. Of the above datasets, we chose Choral Singing Dataset and ESMUC Choir Dataset for training, validation and testing, while Cantoria Dataset was used for demonstration purposes when presenting our findings. This is because the former two datasets had 16 and 12 singers respectively, distributed between SATB sections, which allowed for data augmentation to be carried out.

4.3   Data Augmentation

Before being employed in the corresponding models, some measures had to be taken to assure consistency because the two datasets used for training had some minor differences. For example, the recordings for the ESMUC Choir Dataset had to be transformed from the standard 22KHz sample rate to 44KHz sample rate [2,3]. In addition, the other models needed to be arranged according to song, which meant that each of the four voice stems for a certain song needed to go into the same folder as the mixture that included all of the voice recordings.

To address the lack of a comprehensively annotated dataset, data augmentation was applied to the three training datasets [2,3]. This was possible due to the multi-track nature of the datasets, along with there being multiple singers for each vocal stem. Each song in the dataset is divided into four vocal stems, each containing three to four distinct recordings by different vocalists. Consequently, artificial mixed recordings for each song are generated by combining one singer per vocal SATB sections, resulting in a total of 256 distinct mixes for each segment, given that there are four singers involved.

To maintain consistency across various models trained locally or in the cloud, these procedures are automated using Python scripts executed on the raw dataset. To generate undiscovered mixes for the testing dataset, we excluded one voice stem from each dataset. This reduction significantly decreased the total number of possible combinations. The resulting dataset comprises 687 unique mixes, which were divided into a roughly 80-10-10 split for training, validation, and testing purposes.

5   Experimentation and Results

5.1   Vocal Separation Results and Comparisons

To begin separating vocals in our choir dataset, we first conducted an 80-10-10 split to allocate data for training, validation, and testing purposes. We evaluated two models, Open-Unmix and BSRNN, for dividing mixed choir vocals into different SATB sections.

The evaluation was based on the Signal-to-Distortion Ratio (SDR), a widely used benchmark for source separation competitions [2,25,26]. SDR, measured in decibels (dB), indicates the quality of separation, with a higher SDR indicating clearer distinction of the desired voice from other vocalists or background noise.

At the core of how SDR functions is the assumption that the estimate of a signal source comprises four separate components:

$$\hat{s}_i = s_{target} + e_{interf} + e_{noise} + e_{artif} \tag{1}$$

where s_{target} is the true source; e_{interf}, e_{noise}, and e_{artif} are error terms for interference, noise and artifacts respectively [26].

In our context, the estimate of the source is the mixture track, while the true source refers to the individual recordings of each vocal stem. Interference, or unwanted signals from other sources, would mainly be the voices of other signals, while noise and artifacts are random signals added during measurement and processing. These components are found in the formula for SDR [2,25,26]:

$$SDR := 10log_{10}(\frac{||s_{target}||^2}{||e_{interf} + e_{noise} + e_{artif}||^2}) \tag{2}$$

It can be inferred that SDR is related to the ratio of the target source to the unwanted signals, and can take negative and positive values. A positive value would mean that in the output, the signal of the true source is greater than that of the unwanted signals, while the opposite is true for a negative value. Hence, we aim to maximise the SDR value, which implies that the voice of the desired SATB singer is much more prominent than the other singers and background noise.

We compared our findings with those of Petermann's study [16], which employed State-of-the-Art (SOTA) models for the same choir source separation challenge. For comparison purposes, we selected the top and bottom performing SOTA models based on specific evaluation criteria.

Our model underwent evaluation using the Choral Singing Dataset, which aligns with the dataset utilized in Petermann's study [16]. This dataset choice ensures consistency and enables a direct comparison between our results and those reported by Petermann.

By leveraging the same dataset and evaluating against both the best and the worse performing SOTA models, we aimed to provide a reliable assessment of our findings in relation to existing research in the field of choir source separation.

The Results Leveraging Open-Unmix: The Open-Unmix model was specially modified for choir recordings, with a focus on training separate models for the SATB choir voices. A few notable changes were the lengthening of the training epoch count from 100 to 200, the augmentation of the batch size from 16 to 64, and the rise in sequence time from 6 to 8 s. In addition, the arrangement of the audio channels was changed from stereo to mono in order to enhance the choir recordings.

The model outperformed State-of-the-Art (SOTA) models in Soprano separation, displaying a noteworthy proficiency with an SDR of 4.68 while achieving an average SDR of 2.92. This achievement is due to the sequence window's 8-second duration increase, which successfully captured distinct voices. However, lower frequency problems plagued the model, especially when it came to differentiating Bass voices. This highlights the need for more improvement in the management of lower frequency bands.

The Results Leveraging BSRNN: Since BSRNN is a frequency-domain model, band-split configurations were used to improve its performance when used with SATB choir music. The model underwent modifications involving a decrease in batch size from 8 to 4, as well as a reduction in the epoch range from 100 to 500 to 30 to 50. Resource and computational limitations are the cause of these modifications.

In spite of this, the model came close to the best SOTA model with an average SDR of 2.84. For example, BSRNN outperformed Open-Unmix in the category of voice separation, with an SDR of 2.86, especially in the tenor voice domain. This performance can be attributed to its band-split arrangement, which successfully distinguished choir sections according to variations in frequency.

Table 2. Performance Comparison

Model	SDR				
	Average	*Soprano*	*Alto*	*Tenor*	*Bass*
Open-Unmix	**2.92**	**4.68**	3.12	2.13	1.74
BSRNN	2.84	2.16	3.12	**2.86**	3.21
SOTA (Worst) [2]	−4.26	−7.29	1.05	−15.78	4.98
SOTA (Best) [2]	2.88	1.67	**10.70**	−7.13	**7.42**

Comparison and Further Discussion As outlined in the methodology, Open-Unmix and BSRNN were selected based on their distinct capabilities, with the evaluation focusing on their efficacy in separating choir vocals. In terms of the Bass audio component, neither Open-Unmix nor BSRNN match the performance of previous methods, as illustrated in Table 2.

Open-Unmix demonstrates promise in this regard, achieving an average SDR of 2.92, the highest among the evaluated methods. This result is consistent with previous work, which found Open-Unmix to outperform other models on the same dataset [2]. Interestingly, the highest SDR obtained for Alto and Bass components was also achieved by the Open-Unmix model in the previous study [2].

Even though Open-Unmix exhibits better performance in separating the Soprano component compared to other models, it demonstrates weaker performance in separating the Tenor component compared to BSRNN, which emerges as the superior choice for this particular component.

5.2 Converting Audio to Score Sheet Through MIDI File

In the final step of the proposed method, the separated audio files (e.g., vocal tracks) are converted into MIDI format and then into sheet music. For this task, we utilized music21 and AnthemScore software, comparing their performance to select the superior option for this case study. Our findings revealed that Music21 outperformed AnthemScore for this case. Music21, a Python library tailored for computer-aided musicology, offers a plethora of tools for working with musical data, analysis, and notation. It facilitates the processing of MIDI files, their conversion into music21 objects, and the display of music notation. Leveraging Music21, we transform separated vocal tracks into raw MIDI data, subsequently rendering them into human-readable sheet music.

6 Conclusion, Limitations and Future Works

6.1 Conclusion

In conclusion, this paper proposed DLVS4Audio2Sheet, a novel method designed to address the challenges of transcribing choral music into notated music sheets. DLVS4Audio2Sheet demonstrates promising results in segmenting choral audio into individual vocal sections by leveraging advanced deep learning models such as Open-Unmix and BSRNN for vocal separation. Through rigorous evaluation using these deep learning algorithms, we have assessed the effectiveness of DLVS4Audio2Sheet in producing isolated vocals suitable for notated scoring music conversion. The method's ability to select optimal models for further processing enhances the quality of vocal separation, consequently improving the overall audio-to-music sheet conversion process. This research significantly contributes to the advancement of music technology by thoroughly exploring state-of-the-art models, methodologies, and techniques for converting choral audio into music sheets. Moving forward, DLVS4Audio2Sheet holds promise for facilitating more efficient and accurate transcription of choral music, benefiting music enthusiasts, performers, and composers alike.

6.2   Limitations and Future Works

While DLVS4Audio2Sheet presents a promising approach for converting choral audio into music sheets, there are certain limitations to be considered. Firstly, the effectiveness of the method heavily relies on the quality of the input audio data, including factors such as recording quality and background noise levels. In scenarios where the input audio contains significant noise or overlapping vocal sections, DLVS4Audio2Sheet may encounter challenges in accurately segmenting and isolating individual vocal parts.

Secondly, another limitation of DLVS4Audio2Sheet is the scarcity of training data available for deep learning models. Choral music datasets suitable for training and evaluating vocal separation algorithms are limited. This shortage of data may hinder DLVS4Audio2Sheet's ability to generalize effectively across diverse choral music styles and contexts. Furthermore, the lack of diversity in training data could result in overfitting or biases in the learned representations, thereby limiting the method's robustness and performance on unseen or challenging inputs.

Additionally, while DLVS4Audio2Sheet enhances the audio-to-music sheet conversion process by improving vocal separation quality, it may not fully address all challenges associated with transcribing choral music. Factors such as nuances in musical interpretation, tempo variations, and non-standard notation conventions may still require manual intervention or additional post-processing steps.

Lastly, the performance of DLVS4Audio2Sheet is contingent upon the capabilities and limitations of the deep learning models, Open-Unmix and BSRNN, utilized for vocal separation. While these models have demonstrated effectiveness in various applications, they may still struggle with certain complexities inherent in choral music arrangements, such as dynamic vocal interactions with noises. Considering such limitations, future research could also look into exploring other deep learning techniques. For example, applying various LLMs to this research domains will be an interest topic.

In summary, while DLVS4Audio2Sheet represents a significant advancement in the field of music transcription, it is essential to acknowledge its limitations and continue exploring avenues for improvement, particularly in ensuring scalability and robustness in real-world applications.

Acknowledgment. The authors express their sincere appreciation to the following SMU students for their enthusiastic interest and invaluable contributions to this music analysis-related research: Darryl Soh, Wan Lin Tay, Norman Ng, Zhen Ming Tog, Joel Tan, Yan Yi Sim, Enqi Chan, Eric Li Tong, Thaddeus Lee, and Wei Lun Teo. Their dedication has significantly enriched our work.

References

1. Benetos, E., Dixon, S., Duan, Z., Ewert, S.: Automatic music transcription: an overview. IEEE Signal Process. Mag. **36**(1), 20–30 (2018)
2. Chandna, P., Cuesta, H., Petermann, D., Gómez, E.: A deep-learning based framework for source separation, analysis, and synthesis of choral ensembles. Front. Signal Process. **2**, 808594 (2022)
3. Cuesta, H., Gómez Gutiérrez, E., Martorell Domínguez, A., Loáiciga, F.: Analysis of intonation in unison choir singing. In: Proceedings of the 15th International Conference on Music Perception and Cognition / 10th Triennial Conference of the European Society for the Cognitive Sciences of Music, Graz (Austria), pp. 125–130 (2018)
4. Cuesta, H., M.B., Gómez, E.: Multiple f0 estimation in vocal ensembles using convolutional neural networks. arXiv preprint: arXiv:2009.04172 (2020)
5. Cuthbert, M.S., Ariza, C.: music21: a toolkit for computer-aided musicology and symbolic music data (2010)
6. Grais, E.M., Sen, M.U., Erdogan, H.: Deep neural networks for single channel source separation. In: 2014 IEEE International Conference on Acoustics, Speech and Signal Processing (ICASSP), pp. 3734–3738. IEEE (2014)
7. Hershey, J., C.M.: Audio-visual sound separation via hidden Markov models. In: Advances in Neural Information Processing Systems, vol. 14 (2001)
8. Hu, Z., Wang, Z., Ho, S.B., Tan, A.H.: Stock market trend forecasting based on multiple textual features: a deep learning method. In: 2021 IEEE 33rd International Conference on Tools with Artificial Intelligence (ICTAI), pp. 1002–1007. IEEE (2021)
9. Hu, Z., Wang, Z., Wang, Y., Tan, A.H.: MSRL-Net: a multi-level semantic relation-enhanced learning network for aspect-based sentiment analysis. Expert Syst. Appl. **217**, 119492 (2023)
10. Hyvärinen, A., Oja, E.: Independent component analysis: algorithms and applications. Neural Netw. **13**(4–5), 411–430 (2000)
11. Luo, Y., Yu, J.: Music source separation with band-split RNN. IEEE/ACM Trans. Audio, Speech Lang. Process. (2023)
12. Mitsufuji, Y., et al.: Music Demixing challenge 2021. Front. Sign. Process. **1**, 808395 (2022)
13. Ni, J., Young, T., Pandelea, V., Xue, F., Cambria, E.: Recent advances in deep learning based dialogue systems: a systematic survey. Artif. Intell. Rev. **56**(4), 3055–3155 (2023)
14. Nikolsky, A., Alekseyev, E., Alekseev, I., Dyakonova, V.: The overlooked tradition of "personal music" and its place in the evolution of music. Front. Psychol. **10**, 3051 (2020)
15. Parth, Y., Wang, Z.: Extreme learning machine for intent classification of web data. In: Cao, J., Cambria, E., Lendasse, A., Miche, Y., Vong, C. (eds.) Proceedings of ELM-2016. Proceedings in Adaptation, Learning and Optimization, vol. 9, pp. 53–60. Springer, Cham (2018). https://doi.org/10.1007/978-3-319-57421-9_5

16. Petermann, D., Chandna, P., Cuesta, H., Bonada, J., Gómez, E.: Deep learning based source separation applied to choir ensembles. arXiv preprint: arXiv:2008.07645 (2020)

17. Román, M.A., Pertusa, A., Calvo-Zaragoza, J.: Data representations for audio-to-score monophonic music transcription. Expert Syst. Appl. **162**, 113769 (2020)

18. Rosenzweig, S., Cuesta, H., Weiß, C., Scherbaum, F., Gómez, E., Müller, M.: Dagstuhl ChoirSet: a multitrack dataset for MIR research on choral singing. Trans. Int. Soc. Music Inf. Retrieval **3**(1), 98–110 (2020)

19. Schedl, M., et al.: Music information retrieval: recent developments and applications. Found. Trends® Inf. Retrieval **8**, 127-261 (2014)

20. Stöter, F.R., Uhlich, S., Liutkus, A., Mitsufuji, Y.: Open-unmix-a reference implementation for music source separation. J. Open Source Softw. **4**(41), 1667 (2019)

21. Tan, Y.S., Teo, N., Ghe, E., Fong, J., Wang, Z.: Video sentiment analysis for child safety. In: 2023 IEEE International Conference on Data Mining Workshops (ICDMW), pp. 783–790. IEEE (2023)

22. Teo, A., Wang, Z., Pen, H., Subagdja, B., Ho, S.B., Quek, B.K.: Knowledge graph enhanced aspect-based sentiment analysis incorporating external knowledge. In: 2023 IEEE International Conference on Data Mining Workshops (ICDMW), pp. 791–798. IEEE (2023)

23. Thakur, K.K., et al.: Speech enhancement using Open-Unmix music source separation architecture. In: 2022 IEEE Delhi Section Conference (DELCON), pp. 1–6. IEEE (2022)

24. Tzinis, E., Wang, Z., Smaragdis, P.: Sudo RM-RF: efficient networks for universal audio source separation. In: 2020 IEEE 30th International Workshop on Machine Learning for Signal Processing (MLSP), pp. 1–6. IEEE (2020)

25. Tzinis, E., Wisdom, S., Hershey, J.R., Jansen, A., Ellis, D.P.: Improving universal sound separation using sound classification. In: ICASSP 2020-2020 IEEE International Conference on Acoustics, Speech and Signal Processing (ICASSP), pp. 96–100. IEEE (2020)

26. Vincent, E., Gribonval, R., Févotte, C.: Performance measurement in blind audio source separation. IEEE Trans. Audio Speech Lang. Process. **14**(4), 1462–1469 (2006)

27. Wen, Y.W., Ting, C.K.: Recent advances of computational intelligence techniques for composing music. IEEE Trans. Emerg. Top. Comput. Intell. **7**(2), 578–597 (2022)

28. Wu, Y.T., Chen, B., Su, L.: Multi-instrument automatic music transcription with self-attention-based instance segmentation. IEEE/ACM Trans. Audio, Speech Lang. Process. **28**, 2796–2809 (2020)

Explainable AI for Stress and Depression Detection in the Cyberspace and Beyond

Erik Cambria[1(✉)], Balázs Gulyás[1], Joyce S. Pang[1], Nigel V. Marsh[2],
and Mythily Subramaniam[3]

[1] Nanyang Technological University, Singapore, Singapore
`{cambria,balazs.gulyas,joycepang}@ntu.edu.sg`
[2] James Cook University, Singapore, Singapore
`nigel.marsh@jcu.edu.au`
[3] Institute of Mental Health, Singapore, Singapore
`mythily@imh.com.sg`

Abstract. Stress and depression have emerged as prevalent challenges in contemporary society, deeply intertwined with the complexities of modern life. This paper delves into the multifaceted nature of these phenomena, exploring their intricate relationship with various socio-cultural, technological, and environmental factors through the application of neurosymbolic AI to social media content. Through a quantitative and qualitative analysis of results, we elucidate the profound impact of technological advancements on information processing, work culture, and social dynamics, highlighting the role of digital connectivity in exacerbating stressors. Economic pressures and social isolation further compound these challenges, contributing to a pervasive sense of unease and disconnection. Environmental stressors, including climate change, add another layer of complexity, fostering existential concerns about the future. Moreover, the persistent stigma surrounding mental health perpetuates a cycle of silence and suffering, hindering access to support and resources. Addressing these issues necessitates a holistic approach, encompassing societal changes, policy interventions, and individual coping strategies.

Keywords: Stress detection · Depression Detection · Artificial Intelligence · Natural Language Processing · Affective Computing

1 Introduction

In the contemporary world, stress and depression have become pervasive issues, intricately woven into the fabric of modern life. One of the primary culprits is the rapid advancement of technology. While it has undeniably enhanced convenience and connectivity, it has also ushered in a relentless cycle of information overload. We find ourselves inundated with constant streams of data, struggling to sift through the endless notifications and updates bombarding our digital existence. This constant barrage can lead to feelings of overwhelm, anxiety, and a pervasive sense of being constantly "on".

© The Author(s), under exclusive license to Springer Nature Singapore Pte Ltd. 2024
Z. Wang and C. W. Tan (Eds.): PAKDD 2024 Workshops, LNAI 14658, pp. 108–120, 2024.
https://doi.org/10.1007/978-981-97-2650-9_9

The modern workplace culture exacerbates these pressures. Fueled by a relentless pursuit of productivity, many individuals find themselves caught in a ceaseless cycle of long hours and mounting expectations. The pressure to perform, coupled with the looming specter of job insecurity, fosters an environment ripe for stress and burnout. Economic pressures further compound these challenges. With living costs on the rise and economic uncertainty looming large, many individuals find themselves grappling with the weight of financial strain. The struggle to make ends meet adds another layer of stress, exacerbating feelings of anxiety and hopelessness. Yet, perhaps one of the most insidious aspects of the modern era is the pervasive sense of social isolation. Despite the veneer of connectivity offered by social media, many individuals find themselves grappling with profound feelings of loneliness. The digital age has paradoxically alienated us from genuine human connection, leaving us longing for meaningful relationships amidst a sea of superficial interactions.

Additionally, environmental factors such as pollution and climate change cast a looming shadow over our collective psyche, fostering a sense of existential dread about the future. Coupled with the relentless pressure to measure up to curated versions of perfection portrayed on social media, many individuals find themselves mired in a cycle of comparison and self-doubt. Compounding these challenges is the persistent stigma surrounding mental health. Despite growing awareness, discussions about mental illness are often shrouded in silence and shame. This pervasive stigma can act as a barrier to seeking help, perpetuating a cycle of suffering in silence.

In the context of computer science, early works have used linguistic analysis to detect signs of depression from text. These had two main drawbacks: they were not very accurate and they only detected presence or absence of depression. More recently, advanced AI techniques were used for a more accurate and finer-grained analysis of depression from both text and videos [2,10,11]. This was part of a larger effort to adopt state-of-the-art AI techniques for healthcare [14], with particular focus on mental health [17,18] and suicidal ideation detection [15, 16]. Despite more performant, however, these new algorithms still had a crucial drawback: they were not explainable, which made them virtually useless for clinicians and mental healthcare experts.

In this work, we apply explainable AI (XAI) [4] to the problem of stress and depression detection from social media to provide a unique lens through which researchers and mental health professionals can observe and understand (and possibly even monitor and prevent) mental health trends in real-time across a wide and diverse population. In particular, we collected about 300,000 tweets about stress and depression and employed state-of-the-art neurosymbolic AI tools to gain a deeper, more nuanced understanding of their potential causes and contributing factors in the modern era.

The remainder of this paper is organized as follows: Sect. 2 introduces our data collection methodology; Sect. 3 describes the data analysis approach undertaken; Sect. 4 discusses results; finally, Sect. 5 offers concluding remarks and outlines future work.

2 Data Collection

We collected our stress and depression dataset from Twitter using the 10 most popular hashtags listed below. We used the new Twitter Pro API package (priced at $5,000 per month) for one month between 1st January to 1st February 2024.

- #MentalHealth: This hashtag is widely used to discuss various aspects of mental health, including stress, depression, anxiety, and other related conditions. It encompasses conversations about personal experiences, coping strategies, and advocacy efforts.
- #Depression: This hashtag specifically focuses on discussions surrounding depression, a mood disorder characterized by persistent feelings of sadness, hopelessness, and loss of interest. It is often used to share personal stories, raise awareness, and provide support to those struggling with depression.
- #Anxiety: Anxiety is a common mental health condition characterized by excessive worry, fear, and apprehension. The #Anxiety hashtag is used to share experiences, coping mechanisms, and resources for managing anxiety-related symptoms.
- #Stress: This hashtag is used to discuss the experience of stress, which refers to the body's response to perceived threats or challenges. Discussions under this hashtag include triggers of stress, coping strategies, and the impact of chronic stress on mental and physical health.
- #SelfCare: Self-care involves intentionally taking care of one's physical, emotional, and mental well-being. The #SelfCare hashtag is used to share tips, practices, and experiences related to self-care activities that can help alleviate stress and promote overall wellness.
- #MentalHealthAwareness: This hashtag is used to raise awareness about mental health issues, including stress and depression, and to promote understanding, acceptance, and support for individuals experiencing mental health challenges.
- #EndStigma: Stigma surrounding mental health can create barriers to seeking help and support. The #EndStigma hashtag is used to advocate for ending the discrimination and prejudice associated with mental illness, fostering a more inclusive and supportive society.
- #MentalHealthMatters: This hashtag emphasizes the importance of prioritizing mental health and acknowledging its significance in overall well-being. It is often used to promote conversations, initiatives, and policies aimed at addressing mental health issues such as stress and depression.
- #Wellness: Wellness encompasses various dimensions of health, including physical, mental, emotional, and social well-being. The #Wellness hashtag is used to share tips, resources, and practices that support holistic health and promote stress reduction and resilience.
- #SelfLove: Self-love involves cultivating a positive and compassionate relationship with oneself. The #SelfLove hashtag is used to promote self-acceptance, self-care, and self-compassion, which are important aspects of managing stress and improving mental health.

Fig. 1. Word cloud representing the top keywords in the dataset

Within one month, we collected a total of one million tweets. After pre-processing (e.g., removal of irrelevant tweets, removal of non-English tweets, removal of duplicates, removal of re-tweets, etc.), we were left with roughly one third of it. The exact distribution of tweets with respect to hashtags is illustrated in Table 1. Figure 1 proposes a visual representation of the most significant terms in the collected dataset (after stopword removal), where the size of each word is proportional to its frequency.

Table 1. Distribution of collected tweets with respect to hashtags.

Hashtag	Start Date	End Date	Tweet Count
#MentalHealth	01-01-2024	01-02-2024	98,034
#Depression	01-01-2024	01-02-2024	82,736
#Anxiety	01-01-2024	01-02-2024	70,304
#Stress	01-01-2024	01-02-2024	31,036
#SelfCare	01-01-2024	01-02-2024	6,802
#MentalHealthAwareness	01-01-2024	01-02-2024	5,124
#EndStigma	01-01-2024	01-02-2024	4,293
#MentalHealthMatters	01-01-2024	01-02-2024	2,640
#Wellness	01-01-2024	01-02-2024	2,479
#SelfLove	01-01-2024	01-02-2024	1,935

Total: 305,383

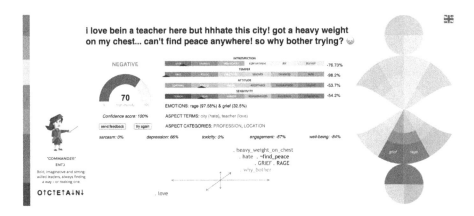

Fig. 2. Sentic API user interface sample.

3   Data Analysis

In order to gain insights from the collected data, we leverage sentiment analysis, a natural language processing (NLP) field in which computational methods are used to extract emotions from text. Different AI techniques have been leveraged to improve both accuracy and interpretability of sentiment analysis algorithms, including symbolic AI, subsymbolic AI, and neurosymbolic AI [5]. Besides traditional algorithms [20] focusing on English text, multilingual [21,26] and multimodal [24] sentiment analysis have also attracted increasing attention recently. Typical applications of sentiment analysis include social network analysis [8], finance [28], and healthcare [3]. In this work, we use Sentic APIs[1], a suite of application programming interfaces available in 80 languages, which employ neurosymbolic AI to perform various sentiment analysis tasks in a fully interpretable manner (Fig. 2). A short description of each API and its usage within this work is provided in the next 12 subsections.

3.1   Concept Parsing

This API provides access to Sentic Parser [6], a knowledge-specific concept parser based on SenticNet [7], which leverages both inflectional and derivational morphology for the efficient extraction and generalization of affective multiword expressions from text. In particular, Sentic Parser is a hybrid semantic parser that uses an ensemble of constituency and dependency parsing and a mix of stemming and lemmatization to extract meaningful multiword expressions. We use the API for extracting words and multiword expressions from text in order to better understand what are the key concepts related to stress and depression. As shown in Fig. 2, for example, some of the concepts extracted are `hate`, `why_bother`, and `heavy_weight_on_chest`.

---

[1] https://sentic.net/api.

3.2   Subjectivity Detection

Subjectivity detection is an important NLP task that aims to filter out 'factual' content from data, i.e., objective text that does not contain any opinion. This API leverages a knowledge-sharing-based multitask learning framework powered by a neural tensor network, which consists of a bilinear tensor layer that links different entity vectors [23]. We use the API to classify stress and depression-related tweets as either objective (unopinionated) or subjective (opinionated) but also to handle neutrality, that is, a tweet that is opinionated but neither positive nor negative (ambivalent stance towards the opinion target). All labels come with a confidence score based on how much SenticNet concepts contributed to the classification output. As depicted in Fig. 2, the confidence score of the proposed example is 100%. Finally, the *Subjectivity Detection module* is also responsible for identifying the language of the input, as indicated in the top-right corner of the UI.

3.3   Polarity Classification

Once an opinionated tweet is detected using the *Subjectivity Detection API*, the *Polarity Classification API* further categorizes this tweet as either positive or negative. This is one of the most important APIs we use to understand the stance of tweeters towards stress and depression. It leverages an explainable fine-grained multiclass sentiment analysis method [27], which involves a multi-level modular structure designed to mimic natural language understanding processes, e.g., ambivalence handling process, sentiment strength handling process, etc. As illustrated in Fig. 2, for example, the extracted polarity is NEGATIVE.

3.4   Intensity Ranking

For a finer-grained analysis, we further process stress and depression classified by the *Polarity Classification API* using the *Intensity Ranking API* to infer their degree of negativity (floating-point number between -100 and 0) or positivity (floating-point number between 0 and 100). In particular, the API leverages a stacked ensemble method for predicting sentiment intensity by combining the outputs obtained from several deep learning and classical feature-based models using a multi-layer perceptron network [1]. As shown in Fig. 2, the extracted polarity of the proposed example is 70 (high intensity).

3.5   Emotion Recognition

This API employs the Hourglass of Emotions [25], a biologically-inspired and psychologically-motivated emotion categorization model, that represents affective states both through labels and through four independent but concomitant affective dimensions, which can potentially describe the full range of emotional experiences that are rooted in any of us. We use the API to go beyond polarity and intensity by examining what are the specific emotions elicited by stress and depression in both their ardent supporters and vocal opposers. As depicted in Fig. 2, for example, the emotion spectrum of the input is visualized in terms of the Hourglass Model's affective dimensions, namely: -76.73% Introspection, -98.2% Temper, -53.7% Attitude, and -54.2% Sensitivity. From these, the API also extracts the two top resulting emotion labels, rage and grief, with an intensity of 97.58% and 32.5%, respectively.

3.6   Aspect Extraction

This API uses a meta-based self-training method that leverages both symbolic representations and subsymbolic learning for extracting aspects from text. A teacher model is trained to generate in-domain knowledge, where the generated pseudo-labels are used by a student model for supervised learning [13]. We use the API to better understand stress and depression in terms of subtopics. Instead of simply identifying a polarity associated with the whole tweet, the *Aspect Extraction API* deconstructs input text into a series of specific aspects or features to then associate a polarity to each of them. This is particularly useful to process antithetic tweets, e.g., tweets containing both positive and negative opinion targets. As illustrated in Fig. 2, the aspect terms extracted from the proposed example are `city` and `teacher`, which belong to the aspect categories `LOCATION` and `PROFESSION`, respectively. The UI also displays the affective concepts most relevant to each aspect term (in brackets) which are also colored according to their respective polarities (green for positive and red for negative).

3.7   Personality Prediction

This API uses a novel hard negative sampling strategy for zero-shot personality trait prediction from text using both OCEAN and MBTI models (Fig. 3). In particular, the API leverages an interpretable variational autoencoder sampler, to pair clauses under different relations as positive and hard negative samples, and a contrastive structured constraint, to disperse the paired samples in a semantic vector space [30]. We use the API to study the different personalities and personas involved in stress and depression discussions and, hence, better understand the possible drivers of such discussions. As shown in Fig. 2, for example, the MBTI personality extracted is ENTJ (Extraverted, iNtuitive, Thinking, and Judging) and the OCEAN personality traits extracted are O↑C↑E↑A↓N↓, i.e., high Openness, high Conscientiousness, high Extraversion, low Agreeableness, and low Neuroticism.

3.8   Sarcasm Identification

This API combines commonsense knowledge and semantic similarity detection methods to better detect and process sarcasm in text. It also employs a contrastive learning approach with triplet loss to optimize the spatial distribution of sarcastic and non-sarcastic sample features [29]. We use the API to understand how much stress and depression are subject to satire and critique but also

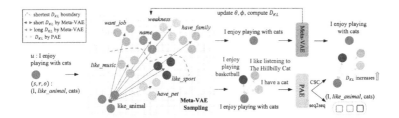

Fig. 3. Personality prediction visualization sample

to increase the accuracy and reliability of the *Polarity Classification API*. As sarcasm often involves expressing a sentiment that is opposite to the intended emotion, in fact, it may lead to polarity misclassification and, hence, generate wrong insights and conclusions. The sarcasm score goes from zero (no sarcasm detected) to 100 (extremely sarcastic content). As depicted in Fig. 2, no sarcasm was detected in the proposed example.

3.9   Depression Categorization

This API employs a novel encoder combining hierarchical attention mechanisms and feed-forward neural networks. To support psycholinguistic studies, the model leverages metaphorical concept mappings as input. Thus, it not only detects depression, but also identifies features of such users' tweets and associated metaphor concept mappings [12]. We use it to discover common causes of depression but also to study different reactions to it by different users. The depression score ranges from zero (no depression detected) to 100 (severe depression). As illustrated in Fig. 2, for example, the depression score is 66%.

3.10   Toxicity Spotting

This API is based on a multichannel convolutional bidirectional gated recurrent unit for detecting toxic comments in a multilabel environment [19]. In particular, the API extracts local features with many filters and different kernel sizes to model input words with long term dependency and then integrates multiple channels with a fully connected layer, normalization layer, and an output layer with a sigmoid activation function for predicting multilabel categories such as 'obscene', 'threat', or 'hate'. The toxicity score goes from zero (no toxicity detected) to 100 (highly toxic content). As shown in Fig. 2, there was no toxicity detected in the proposed example.

3.11   Engagement Measurement

Measuring engagement is important to understand which specific topics or events are more impactful for both stress and depression. This API employs a graph-embedding model that fuses heterogeneous data and metadata for the classification of engagement levels. In particular, the API leverages hybrid fusion methods for combining different types of data in a heterogeneous network by using semantic meta paths to constrain the embeddings [9]. The engagement score ranges from -100 (high disengagement) to 100 (high engagement). As depicted in Fig. 2, for example, the engagement score is -67%.

3.12 Well-Being Assessment

This API leverages a mix of lexicons, embeddings, and pretrained language models for stress detection from social media texts [22]. In particular, the API employs a transformer-based model via transfer learning to capture the nuances of natural language expressions that convey stress in both explicit and implicit manners. The well-being score ranges from -100 (high stress) to 100 (high well-being). As illustrated in Fig. 2, the well-being score is -84% in the proposed example.

4 Results

In this section, we discuss the most important insights gained through the use of Sentic APIs on the collected dataset. The *Concept Parsing API* enabled us to discover what are the current hot topics related to stress and depression, e.g., anxiety, chronic_stress, burnout, sadness, hopelessness, despair, worry, isolation, exhaustion, and social_withdrawal.

Through the *Subjectivity Detection API*, we realized that the vast majority of stress and depression tweets were opinionated. The unopinionated tweets were mostly promotional and advertising posts. This was further validated by the results of the *Intensity Ranking API*, which were high for both negative and positive spectrum.

By processing subjective text using the *Polarity Classification API*, we obtained 59% negative tweets and 41% positive tweets (at least for the time window of our analysis). For the former group, the most common MBTI personality type was ENTJ and the predominant emotion was anxiety. The latter group (the tweeters promoting self-help), instead, was characterized by an INFJ personality trait and a predominant emotion of acceptance.

The *Sarcasm Identification API* has flagged a subtle presence of sarcasm within the context of this particular topic, albeit at a low level. Unlike many other topics discussed on social media, stress and depression do not seem to lend themselves well to sarcasm, probably because of the lack of ambiguity and straightforwardness gravitating around the subject.

As expected, the *Depression Categorization API* observed high levels of depression. Some individuals used Twitter as a means of reaching out for support and validation from their online communities. Some others shared their thoughts and experiences to externalize their internal struggles and potentially receive empathy and support from others. Finally, some users used Twitter as an outlet to vent their frustration, anger, or despair.

The *Toxicity Spotting API* also did not pick up much toxic content. One significant factor is the nature of stress and depression, which inherently lends itself to more neutral or consensus-based discussions, minimizing the potential for conflict or toxicity. Most individuals approached the topic with openness, curiosity, and a willingness to listen to differing perspectives and, hence, fostered an environment conducive to rather constructive dialogues without personal attacks nor hostility.

The *Engagement Measurement API* exhibited high levels of disengagement, most likely caused by emotional exhaustion (characterized by feelings of emptiness, apathy, and detachment) but also anhedonia (the inability to experience pleasure) and social withdrawal.

The *Well-being Assessment API* detected high levels of stress involving emotional intensity, uncertainty, conflict, pressure, personal vulnerability, information overload, lack of control, and negative social dynamics.

Finally, some very useful insights came from the *Aspect Extraction API*, which helped us individuate key causes of stress and depression. We list the 10 most frequent ones below along with a short elucubration on why such aspects emerged from the over 300,000 tweets as the most prominent.

- `Relationship issues`: Problems within intimate relationships or family conflicts can impact mental health and contribute to depressive symptoms.
- `Financial problems`: Financial stress, such as debt, unemployment, or financial instability, can lead to feelings of hopelessness.
- `Social isolation`: Lack of social support and feelings of loneliness can cause depression, as social connections are essential for emotional well-being.
- `Work-life balance`: Difficulty balancing work responsibilities with personal life and self-care can lead to chronic stress and impact mental well-being.
- `Academic pressure`: Students experience stress and depression due to academic demands, performance pressure, or difficulty coping with coursework.
- `Discrimination`: Experiencing discrimination based on race, ethnicity, gender identity, sexual orientation, or other factors can lead to chronic stress.
- `Chronic pain`: Living with chronic health conditions or experiencing persistent pain can be emotionally draining and exacerbate feelings of depression.
- `Trauma`: Past trauma, including physical, emotional, or sexual abuse, can have long-lasting effects on mental health and increase the risk of depression.
- `Media exposure`: Overexposure to negative news, social media comparison, or unrealistic portrayals of success can contribute to feelings of inadequacy.
- `Environmental factors`: Environmental stressors such as pollution, noise, or overcrowding can contribute to chronic stress and impact mental health.

5   Conclusion

Stress and depression have emerged as prevalent challenges in contemporary society, deeply intertwined with the complexities of modern life. This paper delves into the multifaceted nature of these phenomena, exploring their intricate relationship with various socio-cultural, technological, and environmental factors through the application of neurosymbolic AI to social media content. Through a quantitative and qualitative analysis of results, we elucidate the profound impact of technological advancements on information processing, work culture, and social dynamics, highlighting the role of digital connectivity in exacerbating stressors. Economic pressures and social isolation further compound these challenges, contributing to a pervasive sense of unease and disconnection.

Environmental stressors, including climate change, add another layer of complexity, fostering existential concerns about the future. Moreover, the persistent stigma surrounding mental health perpetuates a cycle of silence and suffering, hindering access to support and resources. Addressing these issues necessitates a holistic approach, encompassing societal changes, policy interventions, and individual coping strategies. By fostering greater awareness, empathy, and collective action, we can strive towards a more resilient and compassionate society, better equipped to navigate the complexities of the modern era.

Acknowledgments. This research/project is supported by the Ministry of Education, Singapore under its MOE Academic Research Fund Tier 2 (STEM RIE2025 Award MOE-T2EP20123-0005).

References

1. Akhtar, M.S., Ekbal, A., Cambria, E.: How intense are you? Predicting intensities of emotions and sentiments using stacked ensemble. IEEE Comput. Intell. Mag. **15**(1), 64–75 (2020)
2. Ansari, L., Ji, S., Chen, Q., Cambria, E.: Ensemble hybrid learning methods for automated depression detection. IEEE Trans. Comput. Soc. Syst. **10**(1), 211–219 (2023)
3. Cambria, E., Benson, T., Eckl, C., Hussain, A.: Sentic PROMs: application of sentic computing to the development of a novel unified framework for measuring health-care quality. Expert Syst. Appl. **39**(12), 10533–10543 (2012)
4. Cambria, E., Malandri, L., Mercorio, F., Mezzanzanica, M., Nobani, N.: A survey on XAI and natural language explanations. Inf. Process. Manage. **60**, 103111 (2023)
5. Cambria, E., Mao, R., Chen, M., Wang, Z., Ho, S.B.: Seven pillars for the future of artificial intelligence. IEEE Intell. Syst. **38**(6), 62–69 (2023)
6. Cambria, E., Mao, R., Han, S., Liu, Q.: Sentic parser: a graph-based approach to concept extraction for sentiment analysis. In: Proceedings of ICDM Workshops, pp. 413–420 (2022)
7. Cambria, E., Zhang, X., Mao, R., Chen, M., Kwok, K.: SenticNet 8: fusing emotion AI and commonsense AI for interpretable, trustworthy, and explainable affective computing. In: Proceedings of the 26th International Conference on Human-computer Interaction (2024)
8. Cavallari, S., Cambria, E., Cai, H., Chang, K., Zheng, V.: Embedding both finite and infinite communities on graph. IEEE Comput. Intell. Mag. **14**(3), 39–50 (2019)
9. Chaturvedi, I., Thapa, K., Cavallari, S., Cambria, E., Welsch, R.E.: Predicting video engagement using heterogeneous DeepWalk. Neurocomputing **465**, 228–237 (2021)
10. Chen, Q., Chaturvedi, I., Ji, S., Cambria, E.: Sequential fusion of facial appearance and dynamics for depression recognition. Pattern Recogn. Lett. **150**, 115–121 (2021)

11. Chiong, R., Budhi, G., Cambria, E.: Detecting signs of depression using social media texts through an ensemble of ensemble classifiers. IEEE Trans. Affect. Comput. **15** (2024)
12. Han, S., Mao, R., Cambria, E.: Hierarchical attention network for explainable depression detection on twitter aided by metaphor concept mappings. In: Proceedings of COLING, pp. 94–104 (2022)
13. He, K., Mao, R., Gong, T., Li, C., Cambria, E.: Meta-based self-training and re-weighting for aspect-based sentiment analysis. IEEE Trans. Affect. Comput. **15** (2024)
14. He, K., et al.: A survey of large language models for healthcare: From data, technology, and applications to accountability and ethics. arXiv preprint arXiv:2310.05694 (2024)
15. Ji, S., Li, X., Huang, Z., Cambria, E.: Suicidal ideation and mental disorder detection with attentive relation networks. Neural Comput. Appl. **34**, 10309–10319 (2022)
16. Ji, S., Pan, S., Li, X., Cambria, E., Long, G., Huang, Z.: Suicidal ideation detection: a review of machine learning methods and applications. IEEE Trans. Comput. Soc. Syst. **8**(1), 214–226 (2021)
17. Ji, S., Zhang, T., Yang, K., Ananiadou, S., Cambria, E.: Rethinking large language models in mental health applications. arXiv preprint arXiv:2311.11267 (2024)
18. Ji, S., Zhang, T., Yang, K., Ananiadou, S., Cambria, E., Tiedemann, J.: Domain-specific continued pretraining of language models for capturing long context in mental health. arXiv preprint arXiv:2304.10447 (2024)
19. Kumar, J.A., Abirami, S., Trueman, T.E., Cambria, E.: Comment toxicity detection via a multichannel convolutional bidirectional gated recurrent unit. Neurocomputing **441**, 272–278 (2021)
20. Nguyen, H.T., Duong, P.H., Cambria, E.: Learning short-text semantic similarity with word embeddings and external knowledge sources. Knowl. Based Syst. **182**, 104842 (2019)
21. Peng, H., Ma, Y., Poria, S., Li, Y., Cambria, E.: Phonetic-enriched text representation for Chinese sentiment analysis with reinforcement learning. Inf. Fusion **70**, 88–99 (2021)
22. Rastogi, A., Liu, Q., Cambria, E.: Stress detection from social media articles: new dataset benchmark and analytical study. In: IJCNN (2022)
23. Satapathy, R., Rajesh Pardeshi, S., Cambria, E.: Polarity and subjectivity detection with multitask learning and BERT embedding. Future Internet **14**(7), 191 (2022)
24. Stappen, L., Baird, A., Cambria, E., Schuller, B.: Sentiment analysis and topic recognition in video transcriptions. IEEE Intell. Syst. **36**(2), 88–95 (2021)
25. Susanto, Y., Livingstone, A., Ng, B.C., Cambria, E.: The hourglass model revisited. IEEE Intell. Syst. **35**(5), 96–102 (2020)
26. Vilares, D., Peng, H., Satapathy, R., Cambria, E.: BabelSenticNet: a commonsense reasoning framework for multilingual sentiment analysis. In: IEEE SSCI, pp. 1292–1298 (2018)
27. Wang, Z., Hu, Z., Ho, S.B., Cambria, E., Tan, A.H.: MiMuSA–mimicking human language understanding for fine-grained multi-class sentiment analysis. Neural Comput. Appl. **35**(21), 15907–15921 (2023)
28. Xing, F., Cambria, E., Welsch, R.: Intelligent asset allocation via market sentiment views. IEEE Comput. Intell. Mag. **13**(4), 25–34 (2018)

29. Yue, T., Mao, R., Wang, H., Hu, Z., Cambria, E.: KnowleNet: knowledge fusion network for multimodal sarcasm detection. Inf. Fusion **100**, 101921 (2023)
30. Zhu, L., Li, W., Mao, R., Pandelea, V., Cambria, E.: PAED: zero-shot persona attribute extraction in dialogues. In: ACL, pp. 9771–9787 (2023)

International Workshop on Temporal Analytics (IWTA 2024)

Finding Foundation Models for Time Series Classification with a PreText Task

Ali Ismail-Fawaz[1]([✉]), Maxime Devanne[1], Stefano Berretti[2], Jonathan Weber[1], and Germain Forestier[1,3]

[1] IRIMAS, Universite de Haute-Alsace, Mulhouse, France
{ali-el-hadi.ismail-fawaz,maxime.devanne,jonathan.weber,
germain.forestier}@uha.fr
[2] MICC, University of Florence, Florence, Italy
stefano.berretti@unifi.it
[3] DSAI, Monash University, Melbourne, Australia
germain.forestier@monash.edu

Abstract. Over the past decade, Time Series Classification (TSC) has gained an increasing attention. While various methods were explored, deep learning – particularly through Convolutional Neural Networks (CNNs) –stands out as an effective approach. However, due to the limited availability of training data, defining a foundation model for TSC that overcomes the overfitting problem is still a challenging task. The UCR archive, encompassing a wide spectrum of datasets ranging from motion recognition to ECG-based heart disease detection, serves as a prime example for exploring this issue in diverse TSC scenarios. In this paper, we address the overfitting challenge by introducing pre-trained domain foundation models. A key aspect of our methodology is a novel pretext task that spans multiple datasets. This task is designed to identify the originating dataset of each time series sample, with the goal of creating flexible convolution filters that can be applied across different datasets. The research process consists of two phases: a pre-training phase where the model acquires general features through the pretext task, and a subsequent fine-tuning phase for specific dataset classifications. Our extensive experiments on the UCR archive demonstrate that this pre-training strategy significantly outperforms the conventional training approach without pre-training. This strategy effectively reduces overfitting in small datasets and provides an efficient route for adapting these models to new datasets, thus advancing the capabilities of deep learning in TSC.

Keywords: Time Series Classification · Deep Learning · Pre-Training Deep Learning · Time Series · Convolutional Neural Networks

1 Introduction

Time series are sequences of data points indexed by time, typically obtained by observing a random variable over consistent intervals. These data sequences are

Z. Wang and C. W. Tan (Eds.): PAKDD 2024 Workshops, LNAI 14658, pp. 123–135, 2024.
https://doi.org/10.1007/978-981-97-2650-9_10

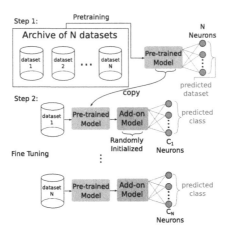

Fig. 1. Summary of the proposed pretext task approach. Given an archive of N datasets, the first step is to train a *pre-trained* model (in blue) on all of the datasets, where the classification task is to predict the dataset each time series belongs to. The second step is to copy the *pre-trained* model and follow it with an *addon* model (in green) randomly initialized. The second step is done independently for each of the N datasets of the archive. After constructing the N new models, they are fine-tuned on each dataset depending on the task of each one.

prevalent in various machine learning applications, including classification [14] and clustering [8], among others. Over the past decade, Time Series Classification (TSC) has witnessed a surge in research activity. This increasing interest spans across diverse fields such as medicine and telecommunications.

Deep learning, with its advanced neural network architectures, offers significant potential for TSC classification [11], often achieving state-of-the-art performance in various TSC tasks. Conventionally, solving a TSC problem with deep learning involves initializing a neural network architecture randomly and feeding it with the training data. However, when the training dataset is limited, this method can lead to overfitting, where the model adapts too closely to the training data, resulting in poor performance on unseen test samples. This challenging problem of having a dataset with few training examples does exist almost everywhere in machine learning research. This common problem reflects a real case scenario and it has been adapted to datasets of the UCR archive, the most comprehensive repository for univariate TSC datasets. This large archive is composed of 128 datasets covering various TSC tasks going from motion recognition to the classification of heart diseases using Electrocardiogram (ECG) signals. The depth of the UCR archive lies in its diverse representation of tasks across multiple domains, often providing several example datasets for each domain.

Gathering additional training samples to address the overfitting issue can be time-consuming and resource-intensive. Furthermore, even if more samples are generated, annotating them typically necessitates expertise, thus introducing additional costs. As a solution, various approaches were proposed in the literature such as data augmentation [9], and the use of hand-crafted generic filters [6]. However, while effective, these methods can introduce noise and disrupt the training process.

To take advantage of having multiple datasets within a given domain, we aim to identify a foundation pre-trained model for each domain of TSC, replacing the random initialization used in traditional techniques. This *pre-trained* foundation model is trained on a shared task among the different datasets. Specifically, the task is to predict the original dataset of each sample. For instance, if we merge two datasets, *dataset1* and *dataset2*, from the same domain, and temporarily disregard their specific target classes, the objective of the pre-trained model becomes discerning the origin of each sample in this combined set.

Once the pre-training phase is completed, the model is fine-tuned for the specific tasks of each dataset. An overview of our proposed methodology is depicted in Fig. 1. After the pre-trained model has been fully trained on the pretext task, the fine tuning stage can follow two different options. The first option is to fine tune the pre-trained model followed by a classification layer with respect to the classification task of the dataset. The second option is to fine tune the pre-trained model cascaded with deeper layers to extract deeper features followed by a classification layer. The first option was followed in the work of [10], where the authors studied the effect of transfer learning on TSC. However, performance was not as good as expected, due to the fact that most target datasets were sensitive on the dataset used as source for the transfer learning.

In this work, we follow the setup of the second option. In particular, we believe that in the first option ignoring deeper meaningful features correlated with one dataset during the fine tuning step implies a strong assumption: the pre-trained model learned the optimal convolution filters that are able to correctly generalize to the classification task. But this may not be the case.

In summary, the main contributions of this work are:

- Novel domain foundation models trained to solve a pretext task to enhance deep learning for TSC;
- Novel Batch Normalization Multiplexer (BNM) layer that controls the multi-dataset (multi-distribution) problem of the batch normalization;
- Extensive experiments on the UCR archive show a significant improvement when using the pre-trained model over the baseline model.

2   Related Work

Many works in the literature have been proposed to address the TSC task and have been evaluated on the UCR archive. These tasks range from similarity based approaches to ensemble models, deep learning, *etc.* In what follows, we present the latest state-of-the-art approaches that addressed the TSC task.

2.1   Deep Learning Techniques

In 2019, the authors of [11] released a detailed review on the latest deep learning approaches for solving TSC on the UCR archive. The two best performing models were Convolutional Neural Networks (CNNs), the Fully Convolutional Network (FCN), and the Residual Network (ResNet) [15]. Moreover, the authors of [12] proposed a new CNN based architecture called InceptionTime, which is an ensemble of multiple Inception models. More recently, new hand-crafted convolution filters were proposed to enhance InceptionTime by [6] with their proposed H-InceptionTime model. It achieves new state-of-the-art performance for deep learners on TSC. Finally, the authors of [4] argued that there is no need for large complex models to solve the TSC task on the UCR archive, but instead they proposed a lighter architecture called LITE. LITE balances between its small number of parameters and its state-of-the-art performance using some boosting techniques.

2.2   Pre-training Deep Learning Techniques

In the last few years, some approaches addressed the TSC task using pre-trained deep learning models. For instance, the work in [10] proposed to apply transfer learning of a deep learning model from a source time series dataset to a target dataset. The deep learning model was trained on a source dataset and then fine tuned on a target dataset. Some works trained a deep learning model with a Self-Supervised task and then used its output features to learn a classifier [7]. The so called "knowledge distillation" is another technique that uses pre-trained models. Following such idea, the authors of [1] used a pre-trained FCN [15] model and distilled its knowledge to a smaller version of FCN. This process helps to balance between a smaller architecture and its performance.

The difference between our proposed approach and the traditional pre-training techniques is the usage of multiple domains during training. It is important to note that the goal of this work is not to solve transfer learning but instead to enhance deep learners when solving direct TSC tasks using a pre-training approach. In what follows, we detail our approach and the used pretext task.

3   Proposed Method

3.1   Pretext Task

A Univariate Time Series (UTS) $\mathbf{x} = \{x_0, x_1, \ldots, x_T\}$ is a vector of T values of a random variable changing with time. Univariate Time Series Classification Dataset (UTSCD) $\mathcal{D} = \{(\mathbf{x}_i, \mathbf{y}_i)\}_{i=1}^{N-1}$ is a set of N UTS with their corresponding label vector $\mathbf{y}$. We denote by C the number of unique labels existing in D. Given a backbone deep learning model for TSC made of n layers, we divided the backbone model into two sub-models. The first sub-model (referred to as the pre-trained model) focuses on learning a pretext task; the latter is an additional randomly initialized model acting as an add-on to the pre-trained model that focuses on the TSC task. The pretext task chosen in this work is the following: given a set of M UTSCD, the task of the pre-trained model is to correctly predict from which

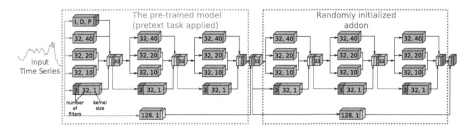

Fig. 2. The H-Inception architecture is divided into two sub-models: the pre-trained model, trained on the pretext task (dotted green rectangle), and the randomly initialized add-on model (dotted red rectangle). The H-Inception model is made of six Inception modules, each module containing three convolutional layers (in orange) and a MaxPooling layer (in magenta), followed by a concatenation (in yellow), a batch normalization layer (in oily), and an activation function (in red). Each Inception module, except the first one, is preceded by a bottleneck layer (in purple) to reduce the dimensionality and so the number of parameters. The first Inception module contains the hybrid addition, which is the hand-crafted convolution filter (in green). Residual connections do exist between the input and the third module, as well as between the third module and the output (in cyan). (Color figure online)

dataset each sample belongs to. It is important to note that one could argue that a more intuitive approach is to combine all datasets and classes and predict a massive class distribution without the need of going through a pretext task. This last approach, however, would result in some issues when no correlation exists between classes of different datasets, so that the class distribution would not have a meaningful representation.

Once the pre-trained model is fully trained, the model is extended by a randomly initialized model. The new constructed model, made of a pre-trained and a randomly initialized sub-model, is then fine tuned on the TSC task for each dataset independently. In summary, the different steps of the whole training procedure are:

- Step 1: Given a set of M UTSCD datasets: $\{\mathcal{D}_0, \mathcal{D}_1, \ldots, \mathcal{D}_{M-1}\}$, where $\mathcal{D}_i = \{(\mathbf{x}_j, \mathbf{y}_j)\}_{j=0}^{N_i-1}$, construct $\mathcal{D}_{PT} = \{(\mathbf{x}_n, \mathbf{yd}_n)\}_{i=0}^{N-1}$, where $N = \sum_{n=0}^{M-1} N_n$, is a dataset including all the time series from $\mathcal{D}_i$ with new labels $\mathbf{yd}$ that represent the dataset the input sample $\mathbf{x}$ belongs to;
- Step 2: Build a pre-trained model, $PT(.)$ with L_{PT} layers trained on $\mathcal{D}$ to correctly classify the dataset each sample belongs to;
- Step 3: Build, for each of the M datasets, a classifier $FT_i(.)$ for $i \in \{0, 1, \ldots, M-1\}$ with $L_{PT} + L_{FT}$ layers;
- Step 4: Fine tune a classifier $FT_i(.)$ for each dataset.

Backbone Model. In this work, we base our model on the-state-of-the-art deep learning model for TSC, the Hybrid Inception architecture (H-Inception) [6]. It is important to note that H-InceptionTime proposed in [6] is an ensemble of five H-Inception models trained with different intializations. For this reason, the backbone architecture in our approach is the H-Inception architecture, and we ensemble the trained models as well following the original work [6,12]. A summarized view of how the H-Inception backbone is decomposed into the pre-trained and fine tuning parts is presented in Fig. 2. Given that the original H-Inception architecture is made of six Inception modules, the first three modules are set to be part of the pre-trained model and the last three are then added to the fine tuning part. We refer to our approach using this specific H-Inception backbone as PHIT (pre-trained H-InceptionTime).

Batch Normalization Multiplexer (BNM). Most deep learning models for TSC [11] that achieve state-of-the-art performance on the UCR archive [2] are convolution-based architectures that also use the Batch Normalization layer with the goal of accelerating the training. In the H-Inception [6] backbone model that we chose, each convolution layer is followed by a Batch Normalization. The role of the Batch Normalization is to learn how to scale and shift the batch samples in order to get a zero mean and unit variance. However, this may be problematic when samples in a same batch are generated from different distributions, *i.e.*, from different datasets, such as in the case of our pre-trained model. For this reason, while training the pre-trained model on the pretext task, multiple Batch Normalization layers should be defined, one for each dataset, so as to replace the one usually used in modern CNN architectures for TSC. For this layer to work, we should then allow the model to connect each sample in the batch to the correct batch normalization layer. A visual representation of the proposed Batch Normalization Multiplexer (BNM) is presented in Fig. 3. From the figure, it can be observed that the BNM takes as input the outcome of the previous layer, with the information of the dataset of the used series, being this information the same one the model is trying to predict. The dataset information goes through the control node of the BNM and chooses which Batch Normalization layer the output node should be connected to.

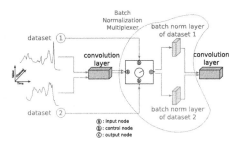

Fig. 3. The proposed BNM, is constituted of multiple batch normalization layers (in oily with blue and red contours) preceded by a multiplexer. This multiplexer has three nodes: (a) the input node, where the input time series goes through, (b) the control node, where the information about the dataset this input time series belong to goes through, and (c) the output node. The path selected for the output node is controlled by the node (b).

4   Results and Analysis

Datasets. To evaluate the performance of our proposed approach, we conducted a series of experiments on the UCR archive dataset [2], which comprises 128 datasets. However, due to redundancies in the archive, our study narrows it down to only 88 datasets. For instance, identical datasets appear multiple times but with varied train-test splits for distinct classification tasks. Such overlaps could compromise the integrity of our model's training as it aims to predict the source dataset of a sample. Moreover, some datasets, while seemingly distinct, merely had varied class counts or were truncated versions of another. A detailed discussion of the reasons for excluding some datasets is reported in Table 1 of the supplementary materials. All datasets underwent a z-normalization prior to training to ensure a zero mean and unit variance. As samples from these datasets may differ in length, zero padding was applied within each batch (rather than before training) to align with the length of the longest series.

Division of the Datasets into Types. The purpose of using a pre-trained model is that of boosting the performance of the deep learning classifier on small datasets using knowledge learned on large ones. This is intuitively most applicable in the case where both the large and small datasets have at least basic information in common. For this reason, we do eight different pretext experiments following the number of dataset types that exist in the UCR archive. In particular, we used all of the datasets of the ECG type to train a pre-trained model, then fine tuned on each dataset independently. These eight types with the corresponding number of datasets are the following: Electrocardiogram (ECG) - 7 datasets, Sensors - 18 datasets, Devices - 9 datasets, Simulation - 8 datasets, Spectrogram - 8 datasets, Motion - 13 datasets, Traffic - 2 datasets, Images contour - 23 datasets.

Implementation Details. The proposed method is implemented in *Tensorflow python* and the code will be available upon acceptance. All of the parameters of the H-Inception model follow the same as in the original work [6]. Each experiment was performed with five different initializations, including the pre-trained and the fine tuned models. Results of multiple runs were assembled together and the model used for evaluation is the best model monitored during training following the training loss. We used a learning rate decay, ReduceLROnPlateau in *keras*, to reduce the learning rate during training by monitoring the train loss with a factor of half. All models were trained on a batch size of 64; the pre-trained model was trained for 750 epochs and the fine tuned model was trained for 750 epochs as well. This last condition ensured us to not train the model for more epochs than the baseline (*i.e.*, the baseline was trained for 1500 in [6]). All experiments were conducted on an Ubuntu 22.04 machine with an NVIDIA GeForece RTX 3090 graphic card with 24 GB of memory.

4.1   Comparing Pre-Training with Baseline (Ensemble)

We present in this section a 1 vs. 1 comparison between our pre-training approach using the H-Inception architecture and the baseline. In what follows, we refer to our approach as Pre-Trained H-InceptionTime (PHIT).

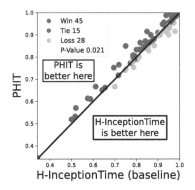

Fig. 4. A 1 vs. 1 scatter plot that compares H-InceptionTime (baseline) and PHIT using the accuracy metric. Each point represents a dataset, where the x and y axis represent the accuracy of H-InceptionTime and PHIT, respectively. A blue point represents a win for PHIT, an orange point a win for H-InceptionTime and a green point a tie. (Color figure online)

Figure 4 represents this 1 vs. 1 comparison by a scatter plot between PHIT and H-InceptionTime. Each point represents a UCR dataset, where the x and y axis report the accuracy metric of H-InceptionTime and PHIT, respectively. The accuracy is evaluated on the test set for each dataset using both methods. This 1 vs. 1 shown that over the 88 datasets, PHIT performs much better than the baseline. From the legend of Fig. 4 it can be seen that PHIT wins 45 times over the baseline; the baseline wins only 28 times. To evaluate the statistical significance of this difference in performance, we presented as well a p-value produced using the Wilcoxon Signed-Rank Test. This p-value, represents the % of confidence of a difference in performance being statistically significant. If the p-value is less than 5% it means there is not enough datasets to conclude a statistical significance in the difference of performance. In this comparison, as seen in Fig. 4, the p-value between PHIT and the baseline is almost 2.1%, which means PHIT significantly outperforms the baseline.

Table 1. The Win/Tie/Loss count between the proposed PHIT approach and the baseline (H-InceptionTime) per dataset domain. The first column presents the number of datasets included per domain followed by the number of Wins for PHIT, number of Ties, and number of Wins for the baseline. We include as well the percentage of number of losses and the average difference in accuracy (PHIT - baseline). A positive value in the last column indicates that on average of all datasets in a specific domain, PHIT performs better than the baseline on the accuracy metric (lowest value 0.0 and highest value 1.0).

Dataset Type	Number of Datasets	Wins of PHIT	Ties of PHIT	Losses of PHIT	Percentage of Losses	**Difference in Average Accuracy (PHIT - Baseline)**
Devices	9	**4**	0	**5**	**55.55%**	+0.0046
ECG	7	**3**	2	2	**28.57%**	+0.0012
Images	23	**14**	2	7	**30.43%**	+0.0087
Motion	13	**11**	1	1	**07.69%**	+0.0179
Sensors	18	**7**	5	6	**33.33%**	+0.0002
Simulation	8	**3**	3	2	**25.00%**	+0.0051
Spectro	8	**3**	2	**3**	**37.50%**	+0.0115
Traffic	2	**0**	0	**2**	**100.0%**	−0.0333

Analysing Performance per Domain. In Table 1, we present a detailed analysis on the performance of the proposed PHIT approach compared to the baseline per dataset domain. We present, for each domain used in the UCR archive, the total number of datasets and the Win/Tie/Loss count with the average difference in performance in the last column. A positive value in the last column confirms that on average PHIT outperforms the baseline on the average accuracy metric. We also present in the 5th column the percentage of number of losses of PHIT. From the table it can be seen that the percentage of losses never exceeds 50% more than twice, and that the average difference in performance is always positive except on one type (*Traffic*). These observations indicate that not only PHIT outperforms the baseline on a global scale of the UCR archive on the majority of domains. This comparison shows that fine tuning a pre-trained model on a generic task, which is in common between multiple datasets is significantly better than the traditional approach.

4.2 Visualizing the Filters

Since we base our work on CNNs, we can compare the space of the learned filters to see the effect of the pre-training approach. In order to visualize this space, we used the t-Distributed Stochastic Neighbor Embedding (t-SNE) visualization technique to reduce the dimensionality of the filters into a 2D plane [6]. By taking the filters of the first Inception module from the baseline, the pre-trained model and the fine tuned model, we can visualize the filters in Fig. 5. In this figure, we consider the experiment over the ECG datasets, where we choose a couple:

ECG200 and NonInvasiveFetalECGThorax1. We chose these two datasets given the difference in size of the training set. For instance, ECG200 has 100 training examples, whereas NonInvasiveFetalECGThorax1 has 1800.

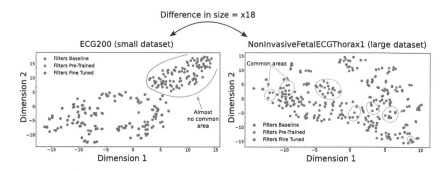

Fig. 5. A two dimensional representation of the filters coming from the first Inception module of the baseline (in blue), pre-trained(red) and fine tuned (green) models. The used datasets in this study are ECG200 (left) and NonInvasiveFetalECGThorax1 (right). The magenta areas represent the areas around the filters of the baseline model. (Color figure online)

From Fig. 5, the filters of the baseline, pre-trained and fine tuned models are presented for each dataset. The first noticeable aspect is that the blue points, representing the filters of the baseline, are quite different from the other red and green points. This ensures that by using the pre-trained model, then fine tuning it, the backpropagation algorithm learns different convolution filters than the traditional baseline approach. The second noticeable thing is that there exists a difference between both plots. On the one hand, in the case of ECG200 (left plot), almost no common areas exist between the filters of the three models. On the other hand, in the case of NonInvasiveFetalECGThorax1 (right plot) there exist many common areas between the filters of different colors. However, there exist some new areas for the pre-trained and fine tuned filters (green and red), which indicates that even though the dataset is large enough, the pre-trained model explored new filters given what it learned from other datasets.

4.3 Comparison with the State-of-the-Art

In what follows, we utilize a comparison technique proposed in [5] called the Multi-Comparison Matrix (MCM). This MCM presents a pairwise comparison between the classifiers as well as their ordering following the average performance. The MCM has shown to be stable to the addition and removal of classifiers, which gives it an advantage over other comparison approaches. The MCM presents as well the Win/Tie/Loss count and a p−value generated using the two tailed Wilcoxon Signed-Ranked Test to study the significance in the difference of performance. The MCM presents as well an ordering of performance of all classifiers following their average performance. In what follows, we present the MCM to compare PHIT to the state-of-the-art approaches including deep and non-deep learning approaches in Fig. 6. It can be concluded that on the 88 datasets of the UCR archive, PHIT outperforms all of the deep learning approaches following the average performance metric. The MCM also shows that given the 88 datasets, no conclusion can be found on the statistical significance difference in performance between PHIT and the state-of-the-art MultiROCKET.

In order to also compare our approach with HIVE-COTE2.0 (HC2) [13] and Hydra+MultiROCKET (HydraMR) [3,14], we only used 86 datasets given that for some datasets of the UCR archive the results are not provided on the original versions for these two models. The scatter plots showing the performance of PHIT compared to HC2 and HydraMR are presented in Fig. 7. On the one hand, this figure shows that PHIT is still not as good as the HydraMR though the scatter plot shows that on 36 datasets, PHIT wins with a significant margin. On the other hand, no conclusion can be made on the statistical significance in the difference of performance between HC2 and PHIT. This concludes that the proposed approach is able to boost a lot the baseline deep learner to achieve HC2 state-of-the-art performance.

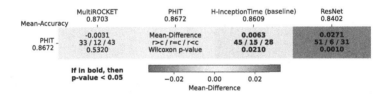

Fig. 6. A Multi-Comparison Matrix (MCM) representing the comparison between the proposed approach PHIT with the state-of-the-art approaches.

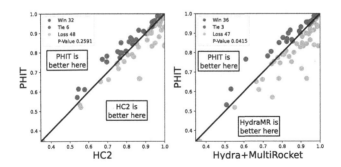

Fig. 7. Two 1 vs. 1 scatter plots representing the comparison between the proposed approach, PHIT, with two state-of-the-art models for TSC, HIVE-COTE2.0 (HC2) and HydraMultiROCKET (HydraMR).

5 Conclusion

In this work, we addressed the Time Series Classification problem by employing innovative pre-trained domain foundation models effectively mitigating overfitting issues in small datasets. Leveraging the UCR archive for evaluation, our methodology involved training models on multiple datasets to accurately classify each sample's original dataset. Subsequent fine-tuning of these models on individual datasets demonstrated superior performance over traditional methods, as evidenced by comprehensive experiments and analyses on the UCR datasets. Our contribution is the creation of domain-specific pre-trained foundation models for time series datasets in the UCR archive, offering a resource for researchers and paving the way for future extensions. This approach, with its inherent generic filters, holds promise for efficient adaptation to new datasets, potentially revolutionizing the training process in time series classification.

Acknowledgment. This work was supported by the ANR DELEGATION project (grant ANR-21-CE23-0014) of the French Agence Nationale de la Recherche. The authors would like to acknowledge the High Performance Computing Center of the University of Strasbourg for supporting this work by providing scientific support and access to computing resources. Part of the computing resources were funded by the Equipex Equip@Meso project (Programme Investissements d'Avenir) and the CPER Alsacalcul/Big Data. The authors would also like to thank the creators and providers of the UCR Archive.

References

1. Ay, E., Devanne, M., Weber, J., Forestier, G.: A study of knowledge distillation in fully convolutional network for time series classification. In: International Joint Conference on Neural Networks (IJCNN) (2022)
2. Dau, H.A., et al.: The UCR time series archive. IEEE/CAA J. Automatica Sinica **6**(6), 1293–1305 (2019)
3. Dempster, A., Schmidt, D.F., Webb, G.I.: Hydra: competing convolutional kernels for fast and accurate time series classification. Data Min. Knowl. Discov. 1–27 (2023)
4. Ismail-Fawaz, A., Devanne, M., Berretti, S., Weber, J., Forestier, G.: Lite: light inception with boosting techniques for time series classification. In: International Conference on Data Science and Advanced Analytics (DSAA) (2023)
5. Ismail-Fawaz, A., et al.: An approach to multiple comparison benchmark evaluations that is stable under manipulation of the compare set. arXiv preprint arXiv:2305.11921 (2023)
6. Ismail-Fawaz, A., Devanne, M., Weber, J., Forestier, G.: Deep learning for time series classification using new hand-crafted convolution filters. In: IEEE International Conference on Big Data (IEEE BigData), pp. 972–981 (2022)
7. Ismail-Fawaz, A., Devanne, M., Weber, J., Forestier, G.: Enhancing time series classification with self-supervised learning. In: International Conference on Agents and Artificial Intelligence (ICAART) (2023)
8. Ismail-Fawaz, A., et al.: ShapeDBA: generating effective time series prototypes using shapeDTW barycenter averaging. In: ECML/PKDD Workshop on Advanced Analytics and Learning on Temporal Data (2023)

9. Ismail Fawaz, H., Forestier, G., Weber, J., Idoumghar, L., Muller, P.A.: Data augmentation using synthetic data for time series classification with deep residual networks. In: ECML/PKDD Workshop on Advanced Analytics and Learning on Temporal Data (2018)
10. Ismail Fawaz, H., Forestier, G., Weber, J., Idoumghar, L., Muller, P.A.: Transfer learning for time series classification. In: IEEE International Conference on Big Data (Big Data) (2018)
11. Ismail Fawaz, H., Forestier, G., Weber, J., Idoumghar, L., Muller, P.A.: Deep learning for time series classification: a review. Data Min. Knowl. Disc. **33**(4), 917–963 (2019)
12. Ismail Fawaz, H., et al.: Inceptiontime: finding AlexNet for time series classification. Data Min. Knowl. Disc. **34**(6), 1936–1962 (2020)
13. Middlehurst, M., Large, J., Flynn, M., Lines, J., Bostrom, A., Bagnall, A.: Hivecote 2.0: a new meta ensemble for time series classification. Mach. Learn. **110**(11-12), 3211–3243 (2021)
14. Middlehurst, M., Schäfer, P., Bagnall, A.: Bake off redux: a review and experimental evaluation of recent time series classification algorithms. arXiv preprint arXiv:2304.13029 (2023)
15. Wang, Z., Yan, W., Oates, T.: Time series classification from scratch with deep neural networks: a strong baseline. In: International Joint Conference on Neural Networks (IJCNN) (2017)

Next Item and Interval Prediction of New Users Using Meta-Learning on Dynamic Network

Jun-Hong Cai, Yi-Hang Tsai, Chia-Ming Chang, and San-Yih Hwang[✉]

National Sun Yat-sen University, Kaohsiung, Taiwan
{m104020041,d094020002,d094020001}@nsysu.edu.tw,
syhwang@mis.nsysu.edu.tw

Abstract. Recommendation systems play a pivotal role in diverse real-world scenarios, offering personalized suggestions to users. Despite their significance, the cold-start problem poses a formidable challenge for both conventional recommendation systems and sequential recommendations. The entry of new users or items into the system inhibits accurate recommendations due to the lack of prior interactions. To tackle this issue, researchers have delved into employing meta-learning techniques. However, predicting the time interval of interactions for new users remains a persistent challenge. This paper presents an innovative approach to forecasting the next item and the associated time interval of new user and item interactions. Our method leverages meta-learning techniques within the context of a dynamic graph structure. It showcases superior performance when compared to previous methods using three benchmark datasets, effectively addressing the cold-start problem. The instructive experiments underscore the efficacy of our proposed method in handling the next item and time interval prediction, thereby contributing to the advancement of sequential recommendation systems.

Keywords: Sequential recommendation · Graph neural networks · Cold-start · Meta-learning · Timing predictions

1 Introduction

Sequential recommendation (SR) systems have received increased research interest in recent years. In contrast to traditional recommendation systems (RS) overlooking the temporal dimension of interactions, SR is known for capturing and preserving the dynamic semantics inherent in user-item interactions, particularly in domains with evolving user preferences. Recognizing that the timing of interactions, including factors such as recency, plays a crucial role in forecasting future user behavior, SR systems have garnered attention for their ability to address the temporal dynamics inherent in recommendation scenarios.

In the field of RS, numerous studies have successfully employed Graph Neural Network (GNN) to tackle the task of completing bipartite user-item matrices [7,16,22,24]. Likewise, in the SR field, a line of research applies dynamic

Z. Wang and C. W. Tan (Eds.): PAKDD 2024 Workshops, LNAI 14658, pp. 136–147, 2024.
https://doi.org/10.1007/978-981-97-2650-9_11

graphs where their topology and structure change over time. These methods effectively address the challenges associated with both discrete and continuous graph structures. A discrete network comprises snapshots of graphs at regular intervals, where the sequences of interactions are divided into subsets based on their timestamps [3,19]. On the other hand, continuous graphs directly incorporating time intervals into the aggregator of the model [4,23]. Nevertheless, SR models also need to address the cold-start problem. When new users or items join the system, RS/SR faces difficulties in providing accurate recommendations due to the absence of prior interactions. The scarcity of interactions results in insufficient information to generate precise recommendations. A common approach is to incorporate heterogeneous information or side information to represent user and item embedding [2,13,14,16]. Another line of research is to leverage meta-learning techniques, such as Model-Agnostic Meta-Learning (MAML) [5], to learn optimal initializations from regular users and transfer them to cold-start users, followed by fine-tuning using a small amount of interactions [9,14,18,21,25]. For instance, Mecos [25] addresses the item cold-start problem by utilizing the Long Short-Term Memory (LSTM) model as the main structure and employing matrix-based meta-learning. It calculates the similarity between users' sequences to recommend cold-start items to similar users. MetaDyGNN [21] is a method that combines meta-learning and graph neural networks to predict future links for new nodes in dynamic networks. It demonstrated better performance in predicting the next items for new users by incorporating time embeddings. Moreover, this approach holds promise for broader application in addressing recommendation-related challenges.

However, previous works in SR mainly focus on predicting the next item a user will most likely to interact with. We argue that predicting when users will purchase the next item is as vital as predicting the next item. In this work, our research goal is to anticipate the time interval of user-item interactions for new users. We propose a continuous dynamic graph model that consists of the node encoder and the interval decoder. The node encoder captures the user-item historical interactions in the dynamic graph, and the interval decoder predicts a specific interval that a new user will interact with an item. The remainder of this paper is organized as follows. Section 2 presents the related works. Section 3 outlines the problem definition along with our proposed method. Then, experiments and the their results are reported in Sect. 4. Finally, we conclude the paper in Sect. 5.

2   Related Work

2.1   Graph-Based Sequence Recommendation Model

The SR system takes the historical sequence of interactions, such as clicked or purchased items, as input and predicts the next interacting item of a given user. MA-GNN [15] applies GNN on an item-item graph to capture order-related factors. It divides the sequence into multiple sub-sequences and performs convolutions on each subsequence. GRU is further applied for encoding sub-sequences, enabling it to capture short-term and long-term user interests.

However, this work focuses on only the items' perspective for representing the correlation of users. Other works incorporate bipartite graphs to establish relationships between users and items. For instance, RetaGNN [8] is a fashion model that utilizes IGMC [24] to generate unseen sequences, considering both long-term and short-term user interests. DRL-SRe [3] and HyperRec [19] adopt discrete graphs, such as regular graphs and hypergraphs, to learn item representations. HyperRec demonstrates that the short-term mechanism is more influential in specific e-commerce scenarios. These works either overlook the time intervals between interactions or use coarse-grained snapshots of graphs to handle dynamic changes. DGSR [23] applies a continuous graph structure and leverages the RNN model as an aggregator. It incorporates time and order into the edges, enabling it to model dynamic and collaborative signals. TGSRec [4] introduces time embeddings to the encoder that considers the time factor in integrating dynamic orders. They compute node embeddings and adopt the Temporal Collaborative Transformer to aggregate neighbors and integrate temporal collaborative signals.

2.2   Meta-Learning for Cold-Start Problem

The cold-start problem poses a significant obstacle in recommendation systems, particularly when users or items have limited or zero interactions, leading to a struggle to make accurate recommendations effectively. Addressing the cold-start problem is paramount to improving the effectiveness and usability of recommendation systems. A standard solution to this issue, so-called the few-shot situation, is to train a base model on a diverse range of tasks and then transfer the model to new tasks by fine-tuning its parameters. One prominent optimization-based meta-learning method that follows this strategy is MAML. Subsequently, some approaches, such as MeLU [11] and MetaHIN [14], have been developed to tackle the user cold-start challenge. These approaches treat users as tasks and train the model on regular users to make predictions for cold-start users. MeLU takes user attributes into account as additional inputs. MetaHIN employs a Heterogeneous Network as the base model to handle the cold-start problem. Additionally, some approaches introduce meta-learning to address cold-start problem in sequential recommendation [9,18,21]. For instance, MetaTL [18] employs a naive transition as the base model, where MAML is used to train regular users and test cold-start users. metaCSR [9] has adopted graph-based models as the underlying structures. It combines GNN and MAML to model cold-start users, with the meta-parameters serving as a positional component.

3   Methodology

This section outlines our framework and method for building a predictive model to estimate future user-item interactions. Given the inherent complexity of predicting such an event, we propose a novel approach that involves converting timestamps to time intervals which simplify the query process and enhance the accuracy of our predictions.

3.1   Problem Definition

The Next Interval Prediction. Given a set of triples $(u, i, t-)$, indicating user u interact item i at time $t-$, the next interval prediction problem is to predict the time interval I_i after a given timestamp t that a given user u will interact with a given item i. Here the interval I_i is a random variable with a probability distribution.

Continuous Dynamic Graph. To effectively model temporal data in the context of GNN, a continuous dynamic graph is employed. Specifically, in order to generate the node representation h_i for node i, which could be a user or an item, at timestamp $t-$, the dynamic GNN (DGNN) algorithm aggregates the representation of neighbor nodes of i and concatenates the time representation $t_{i,j}$ associated with interactions between node i and node j. This process is encapsulated in Eq. (1), where $N(i)$ denotes the set of i's neighboring nodes and $AGG(.)$ is an aggregation function.

$$h_i = AGG(concat(h_j, t_{i,j})), j \in N(i) \tag{1}$$

3.2   Model Architecture

In this study, we aim to address the problem of the next item and its interaction interval prediction of new users. Our method was motivated by MetaDyGNN [21] which employs a node encoder g_θ, where θ is the set of learnable parameters of DGNN, that incorporates a time-aware attention mechanism to effectively capture the temporal dynamics and encode the representations of both user and item nodes. We utilize a time intervals decoder e_ϕ, where ϕ is the set of learnable parameters of DGNN, to predict the time intervals for a user and an item. This decoder is specifically designed to leverage the encoded information and generate accurate predictions of time intervals of interactions between users and items.

Node Encoder. Node encoder g_θ serves as an aggregator that combines interactions of users u and items i before time $t-$ using a temporal attention mechanism. This allows the model to effectively capture the temporal semantics and dynamics associated with users and items. We denote node v's representation by h_v^l, where l is l-th layer in GNN. At the initial stage, we apply random initialization to $h_v^0 \in R^d$. Our model uses distinct weights $W_{r,l} \in R^{d \times d}$, where r is either 'user' or 'item', to aggregate different node types at each layer l. This allows us to learn diverse user-item relations by differentiating user-to-item and item-to-user aggregations so as to capture distinct semantics. In order to incorporate time-aware attention into our model, we leverage the concept of time difference weight Φ based on Bochner's theorem [20]. The weight Φ plays a crucial role in capturing the sequential order of neighbors and their relationships with the source node. It enables our model to assign importance to different neighbors based on their temporal positions and establish a comprehensive understanding of the sequential patterns. In Eq. (2), we illustrate how to aggregate information from these neighboring nodes by node encoder g_θ. We can get the embeddings of node v in layer l at timestamp t as follows,

$$h_{v,t}^l = g_\theta(v,t) = \sigma(\sum_{j\in N(v)} a_{v,j}^l(h_{j,t_{v,j}}^{l-1}||\Phi(t-t_{v,j}))W_{r,l}), \tag{2}$$

$$a_{v,j}^l = softmax(q_{v,j}^l) = \frac{exp(q_{v,j}^l)}{\sum_{k\in N(v)} \exp(q_{v,k}^l)}, \tag{3}$$

$$q_{v,j}^l = \vec{a}_{r,l}(h_v^0||\Phi(0)||h_{j,t}^{l-1}||\Phi(t-t_{v,j})), \tag{4}$$

where θ represents the parameters of g_θ. Temporal information, based on the timestamp $t_{v,j}$ at which nodes v and j most recently interact, is integrated by employing a time embedding encoder Φ and weighting the node transformation with parameter $W_{r,l}$. $||$ means concatenation for the previous layer's node embedding and difference embedding. The model calculates attention scores a using the importance vector q and signify each neighbor's contribution to the final embedding. In Eq. (4) We form the vector q by concatenating the representation of node v, the time embedding for no time difference at the initial stage $\Phi(0)$, the representation of neighboring node j, and the encoded time difference between t and v and j's interaction $t_{v,j}$. We then apply layer-specific weights $\vec{a}_{r,l}$ for each neighbor type r to yield the vector q. After obtaining the node embeddings, we can proceed to predict the next intervals.

Interval Decoder. Interval decoder e_ϕ is a function to predict the time interval during which the first hit will occur. We assume the time interval of the next interaction is a random variable possessing an exponential distribution, because it well captures the user behavior, e.g., next click/purchase, from our preliminary experiments. In our proposed approach, we aim to predict the parameter λ of the exponential distribution by considering the embeddings of the user u, the item i, and the timestamp $t-$. This allows us to calculate the probability of interaction occurring between u and i some time after $t-$:

$$\lambda_{u,i,t-} = e_\phi(h_{u,t-}, h_{i,t-}) = \sigma(MLP(h_{u,t-}||h_{i,t-})), \tag{5}$$

where σ is an activation function that uses the *relu* function, MLP is a multi-layer neural network , and $||$ denotes the concatenation operation. By modeling the probability distribution of these interactions, we can effectively predict when the next interaction between u and i will occur. Figure 1 shows an example for predicting the time interval after t_{15} when user u_{10} and item i_{25} will interact.

Meta-Learning. Our implementation adopts a meta-learning approach with inner and outer loops borrowing from MAML++ [1], an improved version of MAML. The inner loop fine-tunes meta parameters on the new user's support set to find adaptive, user-specific parameters. The outer loop uses these updated

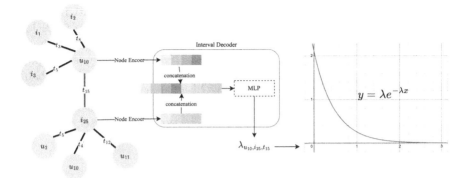

Fig. 1. An example that shows how to determine the time interval between user and item interaction after a specific time. Here to predict when user u_{10} will interact with item i_{25} after time t_{15}, our model aggregates u_{10} and i_{25}'s neighbors before t_{15}. The decoder uses these embeddings to determine the time interval distribution, modeled as an exponential distribution with λ as the parameter.

parameters to perform inference on the new user's query set. This process enables us to adopt the model to individual users. Our model optimizes two parameter sets — node encoder θ and interval decoder ϕ — jointly as a unified set ψ. Computing gradients and updating weights involves a single backward propagation on ψ. Jointly optimizing θ and ϕ enables end-to-end training of all model components. In the inner loop of our algorithm, we leverage the support set to update the initialization of the meta parameters as follows:

$$\psi_u \leftarrow \psi - \beta_{inn} \nabla_\psi \mathcal{L}(S_u, \psi),$$

where $\mathcal{L}(S_u, \psi)$ is the loss incurred by the support set S_u of user u after forwarding in the inner loop using the meta parameters. We use the per-layer, per-step learning rates from MAML++ for efficient learning. This allows dynamically determining distinct learning rates β_{inn} for each step and layer when updating ψ. It enables more fine-grained optimization. We also train individual outer loop learning rates β_{out} to provide adaptability and mitigate overfitting.

Upon completing the fine-tuning process for the individual users, we obtain the losses of the query sets using the adaptive weights. We can update the meta parameters as follows:

$$\psi \leftarrow \psi - \beta_{out} \sum_{u=1}^{U} \sum_{n=1}^{N} c_n \mathcal{L}(Q_u, \psi_u),$$

where U is the total number of users. Q_u represent the user u's corresponding query set. ψ_u denotes the adaptive weights for user u. Furthermore, we calculate the loss of the query set for each step of the inner loop, resulting in N losses for a query set. c_n denotes the importance weight of the query set loss at step n, which is used to compute the weighted sum. The training process is same to metaDyGNN.

Loss Computation. We calculate the parameter λ of the exponential distribution and then use it to compute the cumulative distribution function (CDF) for event probability within a given time interval. The CDF of an exponentially distributed random variable X is $P(X \leq x) = 1 - e^{-\lambda x}$. For instance, let temporal granularities be days, with λ estimated from the Interval decoder, the probability of an event occurring within a day is $1 - e^{-\lambda}$ and after a day is $e^{-\lambda}$. Similarly, the likelihood of an event within two days is $1 - e^{-2\lambda}$, and after two days is $e^{-2\lambda}$. We formally define the loss function for the interaction between user u and item i after $t-$ as follows:

$$loss_{u,i,t-} = \frac{1}{\Omega} \sum_{j=1}^{\Omega} BCE(\hat{y}_{u,i,t-,j}, y_{u,i,t-,j}),$$

where $\hat{y}_{u,i,t-,j}$ denote the predicted interaction probability between user u and item i within the j-th time intervals after t. BCE is the binary cross entropy and Ω is number of time interval in evaluation setting. Meanwhile, $y_{u,i,t-,j}$ represents the ground truth which is 1 when an interaction occurs within the j-th interval, and 0 otherwise For meta-learning, we treat each user as a unique task and sample k instances with positive and negative labels per user, refining model parameters via support/query sets. The inner loop trains a user-specific base model with the support set, and the outer loop optimizes the base model's initialization with the query set. We determine the task's loss for both loops as follows:

$$\mathcal{L}_u = \frac{1}{2k} \sum_{j=1}^{2k} loss_{u,i_j,t-j},$$

where $2k$ represents the total number of positive and negative shots.

4   Evaluations

As there is no prior work on next interval prediction, we compared our model to static graph models based on prediction accuracy and AUC. We then consider our model's performance on sequential recommendation for next item prediction.

4.1   Experimental Setup

Datasets. In our experiment, we incorporate three benchmark datasets: Amazon movie reviews, Yelp restaurant reviews, and the Goodreads book review dataset. These datasets span diverse domains and user preferences, enabling us to assess our model's performance across varied recommendation scenarios. Furthermore, these datasets feature new users with limited interactions, providing a comprehensive test of our model's adaptability and accuracy. To focus on relevant interactions, we filter out early sparse interactions until the frequency increases significantly. Specifically, we truncate the timeline to a point where the frequency of interactions becomes significantly higher. We show the statistics of our datasets in Table 1.

Table 1. Statistics of datasets

	Amazon Movie	Yelp	Goodread
#Users	427,788	279,244	348,082
#Items	121,909	141,286	145,006
#Interactions	3,083,138	3,307,899	11,076,125

Baseline Models. To evaluate our model's effectiveness for the next interval prediction, we compare our model with static graph models that do not model the dynamic preferences of users such as GraphSage [6] and GAT [17]. For next item prediction, we compare our model with existing sequential recommendation models, including SASRec [10], TiSASRec [12], and DGSR [23] This assesses our model against existing graph-based sequential approaches.

Evaluation Settings. We divide the dataset into two equal segments based on the timeline: we use the first half for training and allocate the second half for validation and testing. We ensure that the users in the training and testing sets are mutually exclusive, which guarantees that the model does not have prior knowledge of the users in the testing and validation sets. We add 12 h to the timestamp of sampled interactions of the users to create the $t-$ label. For each user, we sample N links from the data pool to create support and query sets. The number of links sampled varies depending on the dataset. For the Amazon dataset, we set $N = 6$, while for the Yelp and Goodreads datasets, we set $N = 8$. For both the next interval prediction and next item prediction tasks, we sample an equal number of negative and positive instances to create balanced datasets. In our model, we generate predictions for the next intervals based on the specific dataset. We set the number of prediction time interval to five for all datasets. For the Amazon dataset, the length of each interval is set to fifteen days. For the Yelp dataset, we set per interval with a length of seven days. For Goodreads dataset, an interval is set as a day due to the abundance of data.

Evaluation Metric. We evaluate the next interval prediction as a binary classification (interact or not) by accuracy (ACC) and AUC. Specifically, for interacting items in the query set of each user in the test dataset, our method can obtain the probability of interaction in the next time interval. By varying the threshold, we may obtain the precision and recall for the user. Thus, AUC serves as the performance metric. Besides, we set the threshold at 0.5 to consider whether the prediction of our model is classified to true or false on ACC metric. For the next item prediction, we rank the items according to the model's output λ and compute the NDCG@5 and HIT@5 scores. Note that we evaluate our model by querying only new unseen users to assess performance on the cold-start problem of modeling previously unobserved user behavior.

Hyperparameter Settings. We consider the 16 most recent neighbors for aggregation in each layer, and we utilize 2 layers for aggregation and modeling of the temporal semantics. The dimension d is set to 256. We randomly drop out 50% of the neighboring nodes during training to enhance model generalizability. For the meta-learning part, we perform 2 inner loop iterations and set number of training epochs to be 20. The learning rate for the inner loop is 0.001, and the learning rate for the outer loop is 0.0001.

4.2   Experimental Result

Table 2. Experimental result of few-shot next interval prediction. Best performance is shown in bold.

Method	Amazon (6-shots)		Yelp (8-shots)		Goodread (8-shots)	
	ACC	AUC	ACC	AUC	ACC	AUC
GraphSAGE	0.5639	0.6088	0.6378	0.6531	0.6165	0.7446
GAT	0.5449	0.7284	0.6139	0.7606	0.5351	0.8326
Our model	**0.6957**	**0.7570**	**0.7319**	**0.7915**	**0.7731**	**0.8608**

Next Interval Prediction. Our model outperforms the static graph models across all three benchmark datasets, as shown in Table 2. The time-aware attention in our GNN architecture enables the modeling of temporal dynamics and time intervals to better capture evolving user patterns and preferences, leading to better prediction accuracy. Furthermore, our model benefits from node and interval decoders that allows us to model interpretable recommendations. The experimental results demonstrate that the decoder takes into account the next time of occurrence of interactions. By utilizing the exponential distribution, our model is able to make more accurate predictions on when a user might interact with a certain item. This aspect of our model enhances its interpretability and provides a clear understanding of the underlying factors influencing the timing of interactions.

Table 3. Experimental result of few-shot next item prediction. Best performance is shown in bold.

Method	Amazon (6-shots)		Yelp (8-shots)		Goodread (8-shots)	
	NDCG@5	Hit@5	NDCG@5	Hit@5	NDCG@5	Hit@5
SASRec	0.2856	0.4052	0.3295	0.4667	0.5793	0.7176
TiSASRec	0.2593	0.3637	0.2884	0.4062	0.511	0.647
DGSR	0.1587	0.2216	0.211	0.3001	**0.7744**	0.8153
Our model	**0.3337**	**0.5712**	**0.4888**	0.6975	0.5893	**0.8414**
Our model(zero-shot)	0.2056	0.4039	0.479	**0.7028**	0.4386	0.681

Next Item Prediction. As indicated in Table 3, our proposed method achieves better or comparable performance for new users across three datasets. By introducing the exponential distribution for interaction interval, our model effectively gives the most effective recommendation for new users, considering their preferences and historical interactions. The derived meta-parameters preserve general prior knowledge, even enabling strong performance in zero-shot setting compared to previous methods. The self-attention-based sequential recommendation baseline models underperform our proposed model as they are not designed for new users with very limited interaction history. Although that TiSASRec incorporates the time information, but it underperfoms SASRec which is essentially the same model without the time features. This might be due to the improper encoding of time features. Our model outperforms DGSR by about 3% on Hit@5 on Goodreads dataset. However DGSR has significantly outperform our model by about 19% on NDCG@5. This might be due to the higher density of the Goodreads dataset, which enables DGSR to better leverage its DGNN for learning prior knowledge.

5   Conclusion

Our study addresses the challenge of predicting the time intervals at which users and items interact in real-world scenarios. We recognize that traditional recommendation algorithms often overlook the importance of timing in their recommendations, and propose a temporal graph approach with time-aware attention to predict time intervals between user-item interactions. Considering these factors, our model can provide more personalized recommendations and recommend items at the right moment, thereby improving the overall user experience. Additionally, our model demonstrates the ability to conquer the cold-start problem, namely the limited user-item interaction history. Our evaluations have shown the effectiveness of our proposed approach in accurately predicting time intervals and making successful recommendations for new users. These results highlight the potential of our model in enhancing recommendation systems by incorporating temporal dynamics through the use of time difference weights and the model's parameters. For future research, it's instructive to investigate the potential of substituting the output of the interval decoder with different probability distributions, which may enable more flexible predictions and make better predictions when facing different scenarios.

References

1. Antoniou, A., Edwards, H., Storkey, A.: How to train your MAML. arXiv preprint arXiv:1810.09502 (2018)
2. Cai, D., Qian, S., Fang, Q., Hu, J., Xu, C.: User cold-start recommendation via inductive heterogeneous graph neural network. ACM Trans. Inf. Syst. **41**(3), 1–27 (2023)
3. Chen, J., Wang, X., Xu, X.: GC-LSTM: graph convolution embedded LSTM for dynamic network link prediction. Appl. Intell. **52**, 7513–7528 (2022)

4. Fan, Z., Liu, Z., Zhang, J., Xiong, Y., Zheng, L., Yu, P.S.: Continuous-time sequential recommendation with temporal graph collaborative transformer. In: Proceedings of the 30th ACM International Conference on Information & Knowledge Management, pp. 433–442 (2021)
5. Finn, C., Abbeel, P., Levine, S.: Model-agnostic meta-learning for fast adaptation of deep networks. In: International Conference on Machine Learning, pp. 1126–1135. PMLR (2017)
6. Hamilton, W., Ying, Z., Leskovec, J.: Inductive representation learning on large graphs. In: Advances in Neural Information Processing Systems, vol. 30 (2017)
7. He, X., Deng, K., Wang, X., Li, Y., Zhang, Y., Wang, M.: LightGCN: simplifying and powering graph convolution network for recommendation. In: Proceedings of the 43rd International ACM SIGIR Conference on Research and Development in Information Retrieval, pp. 639–648 (2020)
8. Hsu, C., Li, C.T.: RetaGNN: relational temporal attentive graph neural networks for holistic sequential recommendation. In: Proceedings of the Web Conference 2021, pp. 2968–2979 (2021)
9. Huang, X., Sang, J., Yu, J., Xu, C.: Learning to learn a cold-start sequential recommender. ACM Trans. Inf. Syst. (TOIS) **40**(2), 1–25 (2022)
10. Kang, W.C., McAuley, J.: Self-attentive sequential recommendation. In: 2018 IEEE International Conference on Data Mining (ICDM), pp. 197–206. IEEE (2018)
11. Lee, H., Im, J., Jang, S., Cho, H., Chung, S.: MeLU: meta-learned user preference estimator for cold-start recommendation. In: Proceedings of the 25th ACM SIGKDD International Conference on Knowledge Discovery & Data Mining, pp. 1073–1082 (2019)
12. Li, J., Wang, Y., McAuley, J.: Time interval aware self-attention for sequential recommendation. In: Proceedings of the 13th International Conference on Web Search and Data Mining, pp. 322–330 (2020)
13. Liu, S., Ounis, I., Macdonald, C., Meng, Z.: A heterogeneous graph neural model for cold-start recommendation. In: Proceedings of the 43rd International ACM SIGIR Conference on Research and Development in Information Retrieval, pp. 2029–2032 (2020)
14. Lu, Y., Fang, Y., Shi, C.: Meta-learning on heterogeneous information networks for cold-start recommendation. In: Proceedings of the 26th ACM SIGKDD International Conference on Knowledge Discovery & Data Mining, pp. 1563–1573 (2020)
15. Ma, C., Ma, L., Zhang, Y., Sun, J., Liu, X., Coates, M.: Memory augmented graph neural networks for sequential recommendation. In: Proceedings of the AAAI Conference on Artificial Intelligence, vol. 34, pp. 5045–5052 (2020)
16. Shi, C., Hu, B., Zhao, W.X., Philip, S.Y.: Heterogeneous information network embedding for recommendation. IEEE Trans. Knowl. Data Eng. **31**(2), 357–370 (2018)
17. Velickovic, P., Cucurull, G., Casanova, A., Romero, A., Lio, P., Bengio, Y.: Graph attention networks. stat **1050**(20), 10–48550 (2017)
18. Wang, J., Ding, K., Caverlee, J.: Sequential recommendation for cold-start users with meta transitional learning. In: Proceedings of the 44th International ACM SIGIR Conference on Research and Development in Information Retrieval, pp. 1783–1787 (2021)
19. Wang, J., Ding, K., Hong, L., Liu, H., Caverlee, J.: Next-item recommendation with sequential hypergraphs. In: Proceedings of the 43rd International ACM SIGIR Conference on Research and Development in Information Retrieval, pp. 1101–1110 (2020)

20. Xu, D., Ruan, C., Korpeoglu, E., Kumar, S., Achan, K.: Inductive representation learning on temporal graphs. arXiv preprint arXiv:2002.07962 (2020)
21. Yang, C., ET AL.: Few-shot link prediction in dynamic networks. In: Proceedings of the Fifteenth ACM International Conference on Web Search and Data Mining, pp. 1245–1255 (2022)
22. Ying, R., He, R., Chen, K., Eksombatchai, P., Hamilton, W.L., Leskovec, J.: Graph convolutional neural networks for web-scale recommender systems. In: Proceedings of the 24th ACM SIGKDD International Conference on Knowledge Discovery & Data Mining, pp. 974–983 (2018)
23. Zhang, M., Wu, S., Yu, X., Liu, Q., Wang, L.: Dynamic graph neural networks for sequential recommendation. IEEE Trans. Knowl. Data Eng. **35**(5), 4741–4753 (2022)
24. Zhang, M., Chen, Y.: Inductive matrix completion based on graph neural networks. arXiv preprint arXiv:1904.12058 (2019)
25. Zheng, Y., Liu, S., Li, Z., Wu, S.: Cold-start sequential recommendation via meta learner. In: Proceedings of the AAAI Conference on Artificial Intelligence, vol. 35, pp. 4706–4713 (2021)

Adaptive Knowledge Sharing in Multi-Task Learning: Insights from Electricity Data Analysis

Yu-Hsiang Chang, Lo Pang-Yun Ting, Wei-Cheng Yin, Ko-Wei Su,
and Kun-Ta Chuang[✉]

Department of Computer Science and Information Engineering, National Cheng Kung
University, Tainan City, Taiwan
{yhchang,lpyting,wcyin,kwsu}@netdb.csie.ncku.edu.tw,
ktchuang@mail.ncku.edu.tw

Abstract. In time-series machine learning, the challenge of obtaining labeled data has spurred interest in using unlabeled data for model training. Current research primarily focuses on deep multi-task learning, emphasizing the hard parameter-sharing approach. Importantly, when correlations between tasks are weak, indiscriminate parameter sharing can lead to learning interference. Consequently, we introduce a novel framework called *DPS*, which separates training into *dependency-learning* and *parameter-sharing* phases. This structure allows the model to manage knowledge sharing between tasks dynamically. Additionally, we introduce a loss function to align neuron functionalities across tasks, addressing learning interference. Through experiments on real-world datasets, we demonstrate the superiority of *DPS* over baselines. Moreover, our results shed light on the impacts of the two designed training phases, validating that *DPS* consistently ensures a degree of learning stability.

1 Introduction

The Internet of Things has become a part of everyday life, impacting areas such as security surveillance and automation of buildings. Its usefulness is especially notable in electric power data, as evidenced by the spread of smart meters in the USA and EU. This has led to vital services such as Non-Intrusive Load Monitoring and Demand Response [11]. Moreover, there's an increasing focus on analyzing Activities of Daily Living (abbreviated as ADL) for enhanced services like remote healthcare [7]. With IoT's growth, deriving insights from sensor data, especially from smart meters, has become more common due to deep learning advancements. While deep learning has shown promise in various domains, especially time series classification, it often requires abundant labeled data. Currently, the process of labeling electricity data is largely done by hand, with some help from computer programs. However, it still requires an expert to validate it, which is often affected by the quality of the data.

Capturing detailed Activities of Daily Living (ADL) labels in electricity data is challenging due to its dynamic nature and diversity influenced by geography, climate, and lifestyle. While crowdsourcing labeling efforts exist [15], they face challenges like privacy concerns and reward systems. In machine learning, the gap between unlabeled and labeled data, particularly in time-series analysis, is significant. This gap under-scores the importance of leveraging large volumes of unlabeled data. Semi-supervised learning, combining labeled and unlabeled data, shows promise for time-series analysis. Additionally, self-supervised learning, a newer unsupervised technique, has achieved success in fields like natural language processing and computer vision. As shown in Fig. 1, this method derives training labels from the data's inherent structure, facilitating the capture of more profound semantic context from the latent structure of unlabeled data [10]. Within the sphere of Multi-Task Learning (MTL), an additional complexity arises known as *negative transfer*, where the enhancement in one task detrimentally affects the performance in another. This issue is particularly pronounced when there's a lack of clear relevance between tasks, often resulting in compromised model effi-ciency and accuracy. Drawing inspiration from this, subsequence prediction has been adopted as the default self-supervised learning task for unlabeled time-series data in [9]. Addressing this challenge, we propose the **D**ynamic **P**arameter **S**haring framework (abbreviated as *DPS*), conceived to tactically guide the learning process through two phases: *dependency-learning* and *knowledge-sharing* phases.

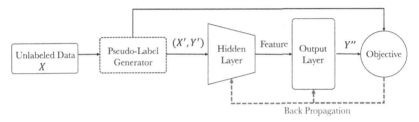

Fig. 1. The process of self-supervised learning.

Currently, many researchers leverage Multi-Task Learning (MTL) [5] to combine these two strategies, thereby enhancing the learning capability and representational quality of models [6,9,12,16]. Our approach begins with a *dependency-learning* phase, during which the model is deeply involved in understanding the unique characteristics of both the main and auxiliary tasks, thereby promoting a holistic understanding. This is followed by a *knowledge-sharing* phase, which adopts a synergistic learning strategy and an advanced alignment technique to coordinate learning across tasks while prevent-ing negative transfer.

However, in most cases, our concern revolves around the performance of a single task, known as the main task. The novelty of the *DPS* framework lies in its fluidity, alternating between these phases to foster an environment conducive to both differenti-ated and integrated learning. This flexibility is instrumental in curbing negative transfer, paving the way for more robust and localized applications in electricity data classifica-tion tasks. Due to factors such as the quantity of labeled data or the objectives of the task itself, certain features might be challenging to learn in the main task but can be easily acquired in another, referred to as the auxiliary task. As such, there's an aspiration to learn these elusive features through the auxiliary task and harness the MTL mechanism to subsequently benefit the main task. For example, in NLP, predicting the presence of positive or negative sentiment words serves as an auxiliary task for sentiment anal-ysis [17]. Similarly, in autonomous driving, single-task models might overlook lane markings as they only occupy a minor portion of the image and are not always present. Hence, "predicting lane markings" can be designated as an auxiliary task, forcing the model to learn this specific feature [5]. Inspired by this, utilizing self-supervised learning as an auxiliary task can offer a potent alternative supervisory signal for feature learning. Especially when learned in tandem with the main task of time-series classification, it can bolster classifier performance in a semi-supervised setting [9].

Given the context, numerous related studies have sought to design auxiliary tasks that provide additional supervisory signals to assist with time-series classification [6,9,12,16]. In the realm of multi-task learning, and especially within the deep multi-task learning framework, hard parameter sharing has emerged as the preeminent approach in neural networks, tracing its origins back to Caruana's research in 1993 [4]. This technique typically entails sharing hidden layers across all tasks while retaining multiple task-specific output layers. We illustrate our methodology using electricity data classification tasks. In our experiments, we demonstrate that our framework can inhibit tasks from interfering with each other's learning in scenarios where inter-task relevance is ambiguous or low. The framework adapts parameter sharing dynamically and avoids negative transfer. In summary, our contributions are as follows:

– The proposed *DPS* framework, a novel strategy in deep multi-task learning, addresses the mitigation of negative transfer by adaptively controlling knowledge sharing among tasks, thereby enhancing task-specific model performance.
– The designed loss function aligns the functionalities of neurons across tasks, enhancing the sharing of knowledge while still allowing for individual task learning.
– Experiments conducted on real-world electrical datasets revealed two key findings. First, *DPS* outperforms baselines in multi-label classification tasks. Second, the two training phases of *DPS* are shown to boost model performance and ensure the stability of model learning.

2   Problem Definition

In this study, we confine our context to the realm of semi-supervised learning. This involves training predominantly on a large volume of unlabeled data while having access to only a limited amount of labeled data. Our aim is to leverage a deep multi-task learning strategy. On one hand, we employ labeled data to train a supervised convolutional neural network for classification, denoted as the main task T_{main}. Concurrently, through self-supervised learning, we capitalize on the vast amount of unlabeled data to extract pertinent features, termed the auxiliary task T_{aux}.

Our objective is to enhance the performance of the classifier through a mechanism known as parameter sharing, or more colloquially, knowledge sharing. We focus specifically on a set of univariate time series samples, $D = D_L \cup D_U$, where $D_L = \{\mathcal{X}_i | \mathcal{X}_i = (x_i^L, y_i^L)\}$ constitutes the labeled dataset, and $D_U = \{\mathcal{X}_j | \mathcal{X}_j = (x_j^U)\}$ represents the unlabeled dataset. For each sample $\mathcal{X}_i = (x_i^L, y_i^L) \in D_L$, there are two pieces of information: (i) an observed time series $x_i^L = \{x_{(i,1)}, ..., x_{(i,t)}\}$ with the length t, and (ii) a set of labels $y_i^L = \{y_1, y_2, ..., y_c | y \in \{0,1\}\}$, where c is the number of all categories, which is the number of electrical appliances in our scenario. On the other side, each sample $\mathcal{X}_j = (x_j^U) \in D_U$ contains an observed time series $x_j^U = \{x_{(j,1)}, ..., x_{(j,t)}\}$

without the corresponding label set. Since the auxiliary task is able to train the model without labels, the samples in the labeled dataset D_L are also adopted as a part of unlabeled dataset D_U ($D_L \subseteq D_U$), which can increase the size of the dataset and help to enhance the performance of the auxiliary task learning.

In this study, we use power consumption data collected from smart meters as an illustrative example. Each sample, denoted as x_i, represents a time series datum corresponding to the i-th time window. This datum signifies the total electricity consumption of a household as recorded by a smart meter. For each of these samples, there's a corresponding true label $y_i = \{y_1, y_2, ..., y_c | y \in \{0,1\}\}$, which is a multi-label set to represent the on/off states of various appliances during that particular i-th time window. Our proposed *DPS* framework aims to learn the main task T_{main} using the labeled dataset D_L. Simultaneously, it learns the auxiliary task T_{aux} with the entire dataset D in a self-supervised manner. By leveraging the knowledge gained from T_{aux}, the framework can effectively predict the multi-labels set y_i^U for each electricity time series $x_i^U \in D_U$.

3   Framework

The proposed DPS framework (an abbreviation for **D**ynamic **P**arameter **S**haring framework) is depicted in Fig. 2. This framework simultaneously addresses two tasks: the main task, a multi-label time series classification T_{main}, which predicts the on-off states of electrical appliances, and an auxiliary task, T_{aux}, aimed at forecasting power consumption over a specified period.

The DPS framework segments the entire model training into two distinct phases: the *dependency-learning* phase and the *knowledge-sharing* phase. In the *dependency-learning* phase, the model undergoes joint optimization for both tasks, with both convolutional and fully-connected layers actively learning. The network parameters are updated to accommodate the demands of each task. To address the issue of negative transfer, as mentioned in Sect.1, the training outcomes from both tasks are integrated with a specifically designed alignment modeling during the *knowledge-sharing* phase. This ensures that neurons in corresponding layers across tasks evolve in closely aligned directions, thus mitigating the issue of negative transfer. The model alternates between the *dependency-learning* phase and the *knowledge-sharing* phase until it converges.

3.1   Tasks Learning Design

Network Architecture. In the proposed DPS framework, we utilize a deep convolutional neural network as the primary architecture. The backbone, denoted by $f_\theta(\cdot)$, as illustrated in Fig. 2, is constructed from multiple layers of one-dimensional convolutional neural networks (1D-CNNs) and sharing layers. The convolutional layers are designed to learn the time series dependency features, capturing the sequential patterns inherent in the data. On the other hand, the sharing layers serve as a medium for exchanging knowledge learned across different tasks. This architecture ensures that the model not only understands the temporal dependencies of the data but also leverages shared knowledge across tasks for improved performance. Upon extracting the features using average pooling (AvgPool), they are directed to the task-specific heads. Both the classification head $h_\mu(\cdot)$ and the prediction head $h_\rho(\cdot)$ are composed of a single fully-connected neural network layer.

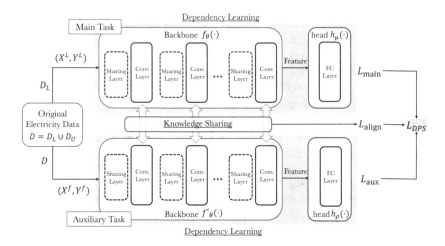

Fig. 2. The overview of the DPS framework.

Self-supervised Tasks. A vast amount of research focuses on devising self-supervised methods. The goal is to extract rich latent features from unlabeled data to enhance the performance of main tasks [12, 16]. Nonetheless, most recent works put little effort into exploring inter-task relevancy [18], which could result in pre-trained representations that are not optimal for downstream tasks. To address this, our approach emphasizes knowledge sharing between tasks, especially when their correlation is unclear or weak, and takes measures to prevent negative transfer. In the designed framework, we select forecasting as the auxiliary task. It is a well-researched and constructible task, but still complex enough to prevent the model from adopting simplistic shortcuts or deceptive strategies [8]. Therefore, we formulate the observed features X^f and the ground truths Y^f to learn the auxiliary task T_{aux}. This task aims to forecast the future m-step electricity data points based on observed time series. The observed time series set X^f and the ground truth set Y^f are defined as follows:

$$\begin{cases} X^f = \{x_i = (x_{(i,1)}, ..., x_{(i,t)}) | x_i \in D\}. \\ Y^f = \{y_i = (x_{(i,t+1)}, ..., x_{(i,t+m)}) | x_i \in X^f\}. \end{cases} \tag{1}$$

3.2 Two-Phase Training

Within our framework, training is divided into two phases. In the first phase, termed the *dependency-learning* phase, both convolutional and fully-connected layers are employed to learn the temporal dependencies within time-series data. The second phase, known as the *knowledge-sharing* phase, centers around the shared layer we introduced. This layer is responsible for discerning which knowledge is to be shared and to what extent. Based on the outcomes from this sharing layer, corresponding neurons in the same layer across different tasks engage in parameter sharing. These two phases alternate until the model converges.

Dependency-Learning Phase. During this phase, the model adjusts its network parameters based on the discrepancy between its predictions and the true labels, a process known as optimization. In our framework, we employ a joint optimization strategy for training. The goal is to simultaneously optimize loss functions across multiple tasks, enabling the model to learn and improve on these tasks concurrently.

Specifically, each task independently optimizes its network parameters. For the main task, T_{main}, which involves multi-label classification, it can be viewed as executing multiple binary classification problems concurrently. As a result, the loss function is defined using the cross-entropy loss, which is apt for multi-label classification:

$$L_{main} = \frac{1}{|D_L|} \sum_{\mathcal{X}_i=(x_i^L,y_i^L)\in D^L} \frac{-\sum_{k=1}^{c} y_i^L(k) \, \log h_\mu^k\left(f_\theta(x_i^L)\right)}{c}, \tag{2}$$

where D_L is the labeled dataset, x_i^L and y_i^L represent the observed electricity time series and the true multi-label set for the i-th time window, respectively. c denotes the total number of electrical appliances, and $y_i^L(k) = \{0,1\}$ is the true label of the k-th appliance. The functions $h_\mu^k(\cdot)$ and $f_\theta(\cdot)$ serve as the head and backbone to learn the main task T_{main}. Additionally, $h_\mu^k\left(f_\theta(x_i^L)\right)$ provides the predicted multi-label set for the i-th time window.

On the other hand, for the auxiliary task T_{aux}, which is designed for electricity forecasting, the loss function is defined using the mean-square error (MSE) as follows:

$$L_{aux} = \frac{1}{|X^f|} \sum_{x_i \in X^f} \left(y_i - h_\rho\left(f_\theta'(x_i)\right)\right)^2, \tag{3}$$

where X^f is the set of observed electricity time series as defined in Eq. 1, and each $x_i \in X^f$ is an electricity time series. $y_i \in Y^f$ is the true label as defined in Eq. 1. The functions $h_\rho(\cdot)$ and $f_\theta'(\cdot)$ are the head and backbone, respectively, tailored to learn the auxiliary task, T_{aux}. The expression $h_\rho\left(f_\theta'(x_i)\right)$ provides a forecast of the m-step electricity points based on x_i.

Our ultimate goal is to minimize both L_{main} and L_{aux} simultaneously. During this process, each task adjusts its network parameters independently.

Knowledge-Sharing Phase. At each training epoch, after training both the main and auxiliary tasks separately, we implement a knowledge-sharing mechanism for layers of the same depth in the backbones of both tasks.

Let the weight vector learned by a neuron in a backbone for task T (T_{main} or T_{aux}) be $\mathbf{w}_{l,k}^T$, where l represents the l-th sharing layer in the backbone, and k denotes the neuron's position in that layer. Upon applying the designed sharing mechanism, the updated weight vector is derived as follows:

$$\mathbf{p}_{l,k}^T = \sum_{T'\in T} \alpha^{T'} \cdot \mathbf{w}_{l,k}^{T'}, \tag{4}$$

where T is the set of all tasks, which includes the multi-label classification task (main task) and the electricity forecasting task (auxiliary task) in our scenario. $\alpha^{T'}$ is a trainable parameter that determines the contribution of neurons from other tasks during the training process. By using Eq. 4, the weights learned by a particular task's backbone are influenced not only by the loss of its own task but also by the losses of other tasks. This setup ensures mutual influence across different tasks during the learning process.

However, arbitrary and substantial adjustments to neuronal parameters might lead to negative transfer. Such misdirection in optimization can result in confusion during task learning. Consequently, to ensure similar learning trajectories across tasks, we designed a loss function based on the behavior of CNNs. Since each feature map in a CNN is produced by a distinct kernel, if identical electricity data inputs yield similar feature maps for tasks with differing objectives, it indicates that the kernels function analogously. Therefore, we assess neuronal similarities across tasks by computing the cosine similarity between feature maps from their convolutional layers.

Let $\mathbf{F}_{l,m}^{\mathrm{main}}$ and $\mathbf{F}_{l,m}^{\mathrm{aux}}$ represent the m-th feature map generated by the l-th convolutional layer of the main task's backbone and the auxiliary task's backbone, respectively. The similarity between $\mathbf{F}_{l,m}^{\mathrm{main}}$ and $\mathbf{F}_{l,m}^{\mathrm{aux}}$ is designed as follows:

$$\phi_{l,m} = \frac{\mathbf{F}_{l,m}^{\mathrm{main}} \cdot \mathbf{F}_{l,m}^{\mathrm{aux}}}{\left\|\mathbf{F}_{l,m}^{\mathrm{main}}\right\|_2 \times \left\|\mathbf{F}_{l,m}^{\mathrm{aux}}\right\|_2}, \tag{5}$$

where $\|\cdot\|_2$ denotes the L2 norm of feature maps.

Based on the estimated similarity between different tasks at each convolutional layer, we defined an *alignment loss* to ensure that the knowledge acquired from each task follows a similar learning trajectory. Our approach for ensuring this consistency evaluates the average similarity S_l^{main} and the standard deviation S_l^{std} of the feature maps produced by the l-th convolutional layer across different tasks' backbones. Specifically, in our scenario, this entails comparing the main task with auxiliary tasks. By monitoring these metrics, our goal is to maintain aligned and consistent learning trajectories across tasks. The alignment loss is derived as follows:

$$\begin{aligned} L_{\mathrm{align}} &= \frac{1}{\mathcal{L}} \sum_{l=1}^{\mathcal{L}} \psi(S_l^{\mathrm{mean}}) * \psi(S_l^{\mathrm{std}}) \\ &= \frac{1}{\mathcal{L}} \sum_{l=1}^{\mathcal{L}} \psi\left(\frac{\sum_{m=1}^{M} \phi_{l,m}}{M}\right) * \psi\left(\sqrt{\frac{1}{M} \cdot \left(\sum_{m=1}^{M} \phi_{l,m} - \phi_l^{\mathrm{avg}}\right)^2}\right), \end{aligned} \tag{6}$$

where $\mathcal{L}$ is the number of convolutional layers in the designed backbones $f_\theta(\cdot)$ and $f_\theta'(\cdot)$. The function $\psi(\cdot)$ represents a logarithmic function. ϕ_l^{avg} denotes the average similarity of all feature maps in the l-th convolutional layer.

Finally, according to Eq. 2, Eq. 3 and Eq. 6, the total loss the proposed *DPS* framework can be written as follows:

$$L_{DPS} = \underbrace{L_{\mathrm{main}} + L_{\mathrm{aux}}}_{\text{dependency learning}} + \lambda \cdot \underbrace{L_{\mathrm{align}}}_{\text{knowledge sharing}}, \tag{7}$$

where parameter λ controls the contribution of knowledge sharing in the proposed *DPS* framework.

Based on our design, we can enhance the prediction of multi-labels for unlabeled electricity time series by leveraging both the main and auxiliary task learning with dynamic knowledge sharing, while also addressing the issue of negative transfer. Although our current setup involves only one main task and one auxiliary task, our *DPS* design is inherently scalable. This scalability allows us to easily expand to incorporate learning from multiple tasks and facilitate broader knowledge sharing. Thus, our approach has the potential to achieve even greater performance improvements in scenarios with a wider array of tasks.

4 Experiments

We conducted experiments to address three primary questions. First, we sought to determine if the proposed *DPS* framework could surpass the performance of other baseline methods. Second, we examined the effects of varying the contributions between the dependency-learning and knowledge-sharing phases. Third, we investigated how the learning speed fluctuates under various knowledge-sharing frequencies.

4.1 Experimental Setup

Dataset Description. For our experiments, we utilized two real-world electrical datasets. The first dataset, sourced from a private electricity provider, encompasses data from December 2016 to March 2017, a total of approximately three months. This dataset includes detailed electrical consumption data from 70 households, capturing the total power consumption and the respective statuses of various appliance switches, with data points recorded at one-minute intervals. For the second dataset, we utilize the real-world ECO dataset[1] that captures electricity consumption from five households. This dataset comprises house profiles, records of total electricity consumption, and the electrical usage of appliances over eight months (from June 2012 to January 2013). For our multi-label classification experiment, we select a household with diverse appliance information to facilitate a comprehensive investigation.

Table 1. Statistics of applied datasets.

Private Dataset		Public Datast (ECO)	
Num. of TS	305,690	Num. of TS	11,519
Selected Appliances	Ratio (label = 1)	Selected Appliances	Ratio (label = 1)
AC	24.7%	TV	26.1%
Kettle	19.7%	Laptop	21.1%
TV	14.8%	Lamp	8.7%
Heater	5%	Dish Washer	1.6%

In our experimental design, we categorize records of total electricity consumption into *on* and *off* states for all selected electrical appliances, the definition of the *on* state of an appliance is the same as [2]. The electricity data is resampled at 30-min intervals, producing a time-series input of length 30 for the input of the model. If an appliance

---

[1] Please refer to https://vs.inf.ethz.ch/res/show.html?what=eco-data..

switches to the *on* state during a 30-minute time series, the label of this appliance in this time series is set as 1; otherwise, it is set as 0. The statistics for the two datasets are presented in Table 1. They display the number of preprocessed time series (TS), the selected appliances, and for each appliance, the proportion of time series where it is labeled as 1.

Comparative Method. To demonstrate the effectiveness of the proposed method, we compare *DPS* with several baselines. **MTL** [9] is a method that leverages self-supervised learning as an auxiliary task, employing strategies like hard parameter sharing and joint optimization. This framework has been embraced by several subsequent studies. **Base** serves as a basic model without auxiliary tasks, effectively operating as a standalone CNN classifier. For the ablation study, we introduce *DPS* **w/o align**, which is a variant of our *DPS* method excluding the alignment loss component.

Evaluation Settings. In our evaluation, we employ widely-used classification metrics such as accuracy (**Acc.**), precision (**Pre.**), recall (**Rec.**), F0.5-Score (**F0.5**), F1-Score (**F1**), and F2-Score (**F2**). Initially, these metrics assess the performance for detecting each electrical appliance, and then we report the average results across all appliances. Higher metric values indicate superior model effectiveness. We also incorporate the Hamming loss (**Ham.**), commonly used in multi-label classification evaluations. Given $y_{i,j}$ as the true label and $\hat{y}_{i,j}$ as the predicted label of the j-th appliance in the i-th testing time series, Hamming loss is defined as follows:

$$\text{Hamming Loss} = \frac{1}{N \cdot L} \sum_{i=1}^{N} \sum_{j=1}^{L} |y_{ij} - \hat{y}_{ij}|, \tag{8}$$

where N represents the number of testing time series, and L is the total number of selected appliances. A lower Hamming loss value indicates better performance.

4.2 Experimental Results

DPS **Performance.** To answer the first question, we compare *DPS* with aforementioned baselines on two datasets. The comparison is summarized in Table 2 and Table 3. Convincingly, our *DPS* approach outperforms baselines across most evaluation metrics. While the method "Base" boasts the highest recall values in both datasets, our accuracy is unsurpassed in the same datasets. This indicates that *DPS* can effectively discern the on-off states of appliances. Furthermore, it suggests that the likelihood of misclassifying an appliance as being on is lower with *DPS*. It's noteworthy that *DPS* outperforms the method "*DPS* w/o align", which doesn't account for alignment loss, across all evaluation metrics. This underscores the importance of ensuring that the knowledge acquired from each task follows a similar learning trajectory. Additionally, the minimal Hamming loss exhibited by *DPS* in both datasets signifies that *DPS* consistently makes fewer wrong predictions across all labels, highlighting its robustness in multi-label classification tasks.

Importance of Different Training Phases. To answer the second question, we vary the value of λ in Eq. 6 to control the importance of the designed alignment loss. λ is set

Table 2. Comparison of multi-label classification on the private dataset. The best results are highlighted in **bold**.

Method	Ham.($\downarrow$)	Acc. ($\uparrow$)	Pre. ($\uparrow$)	Rec. ($\uparrow$)	F0.5 ($\uparrow$)	F1 ($\uparrow$)	F2 ($\uparrow$)
Base	0.072	0.757	0.754	**0.738**	0.750	0.746	**0.742**
MTL	0.070	0.760	0.790	0.692	0.767	0.738	0.711
DPS w/o align	0.075	0.741	0.814	0.610	0.752	0.698	0.650
DPS (proposed)	**0.064**	**0.777**	**0.815**	0.711	**0.790**	**0.759**	0.730

Table 3. Comparison of multi-label classification on the public dataset (ECO dataset). The best results are highlighted in **bold**.

Method	Ham.($\downarrow$)	Acc. ($\uparrow$)	Pre. ($\uparrow$)	Rec. ($\uparrow$)	F0.5 ($\uparrow$)	F1 ($\uparrow$)	F2 ($\uparrow$)
Base	0.111	0.684	0.601	**0.700**	0.621	**0.646**	**0.675**
MTL	0.123	0.685	0.559	0.682	0.581	0.613	0.650
DPS w/o align	0.088	0.687	0.861	0.402	0.686	0.548	0.468
DPS (proposed)	**0.082**	**0.743**	**0.890**	0.491	**0.763**	0.632	0.548

to values from the set $\{ 10^{-3}, 10^{-2}, 10^{-1}, 10^{0}, 10^{1} \}$, and the results are presented in Fig. 3. To highlight the performance difference across different settings, we omitted the Hamming loss results. As can be observed from Fig. 3, variations in λ do not drastically affect the performance of *DPS*. However, the performance of *DPS* suffers when lambda is set too high or too low. This suggests that focusing solely on either the *dependency-learning* phase or the *knowledge-sharing* phase can have a negative effect on *DPS*. It reveals the importance and necessity of our two-phase training structure of our *DPS*.

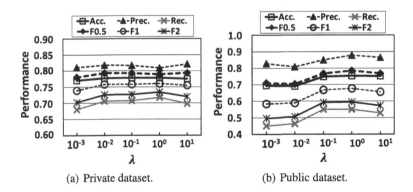

(a) Private dataset.   (b) Public dataset.

Fig. 3. Performance of varying λ.

Effect of Varying Knowledge-Sharing Frequencies. To answer the third question, we examined the effect of knowledge-sharing frequency on the learning speed of *DPS*. Specifically, we fixed the sharing frequency at intervals of every {20, 40, 60, 80, 100} epochs, where knowledge sharing between the main task and the auxiliary task would take place. The results after training *DPS* for 500 epochs are illustrated in Fig. 4. We can observe that the training loss across different settings is fairly consistent in both datasets, suggesting that incorporating the alignment loss in *DPS* ensures a certain level of stability in learning. However, when the sharing frequency is set at every 20 epochs (the red line in Fig. 4(a) and Fig. 4(b)), the rate of decrease in training loss is slower. This suggests that excessive interchange of learned parameters between the main and auxiliary tasks can potentially decelerate the learning speed of *DPS*. This result reveals when the sharing frequency surpasses a certain threshold, it allows for an effective exchange of information learned from both the main and auxiliary tasks.

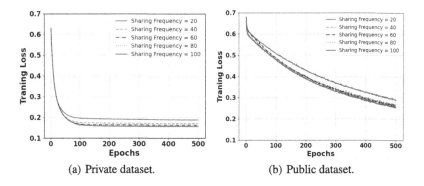

(a) Private dataset.   (b) Public dataset.

Fig. 4. Training loss with varying knowledge sharing frequencies.

5   Related Works

In the evolving landscape of machine learning, self-supervised learning and multi-task learning (MTL) present innovative solutions to longstanding challenges. Supervised models, especially prevalent in image classification, depend heavily on labeled datasets. Acquiring such datasets is often hindered by factors like high costs, time constraints, and issues such as domain-specific intricacies or privacy concerns [10]. Self-supervised representation learning counters this by applying predefined actions on data, creating pseudo labels that guide training. An instance is masked learning, which obscures parts of data, tasking the model with its recovery [9]. Such an approach harnesses the vast amounts of unlabeled data, like online images [3]. Its applications extend beyond pre-processing, touching areas like continual learning [13] and reinforcement learning [14].

Conversely, Multi-Task Learning (MTL) seeks to optimize models for specific metrics but often misses valuable insights from related tasks. By sharing representations between tasks, MTL enhances model generalization [5]. In deep MTL, hard parameter sharing, where hidden layers are shared across tasks but have task-specific output layers, is predominant [4]. This method notably reduces overfitting risks, as demonstrated by research indicating lower overfitting risks with shared parameters than task-specific ones [1]. However, MTL isn't without challenges. Feature learning, which aims to discern common features across tasks, can be influenced by outlier tasks, affecting performance. Balancing shared model parameters and handling outlier tasks remains a pressing issue in MTL.

6   Conclusions

In this paper, we investigate enhancing multi-task learning performance using labeled and unlabeled datasets, considering inter-task correlations. To address the potential drawbacks of hard parameter-sharing methods across tasks, we introduce the *DPS* framework, which includes a *dependency-learning* phase and a *knowledge-sharing* phase. This approach promotes individual task learning and knowledge sharing while maintaining model stability. We show that our *DPS* method surpasses other baselines, validate the effectiveness of its training phases, and emphasize the stability of the model.

References

1. Baxter, J.: A Bayesian/information theoretic model of learning to learn via multiple task sampling. Mach. Learn. **28**, 7–39 (1997)
2. Beckel, C., Kleiminger, W., Cicchetti, R., Staake, T., Santini, S.: The eco data set and the performance of non-intrusive load monitoring algorithms. In: Proceedings of the 1st ACM Conference on Embedded Systems for Energy-Efficient Buildings (2014)
3. Bengio, Y., Lamblin, P., Popovici, D., Larochelle, H.: Greedy layer-wise training of deep networks. In: NIPS (2006)
4. Caruana, R.: Multitask learning: a knowledge-based source of inductive bias. In: International Conference on Machine Learning (1993)
5. Caruana, R.: Multitask learning. Mach. Learn. **28**, 41–75 (1997)
6. Choi, H., Kang, P.: Multi-task self-supervised time-series representation learning. ArXiv: abs/2303.01034 (2023)
7. Devlin, M.A., Hayes, B.P.: Non-intrusive load monitoring and classification of activities of daily living using residential smart meter data. IEEE Trans. Consum. Electron. **65**(3), 339–348 (2019)
8. Gidaris, S., Singh, P., Komodakis, N.: Unsupervised representation learning by predicting image rotations. ArXiv: abs/1803.07728 (2018)
9. Jawed, S., Grabocka, J., Schmidt-Thieme, L.: Self-supervised learning for semi-supervised time series classification. Adv. Knowl. Discov. Data Mining **12084**, 499–511 (2020)
10. Jing, L., Tian, Y.: Self-supervised visual feature learning with deep neural networks: a survey. IEEE Trans. Pattern Anal. Mach. Intell. **43**, 4037–4058 (2019)
11. Kaselimi, M., Protopapadakis, E., Voulodimos, A., Doulamis, N., Doulamis, A.: Towards trustworthy energy disaggregation: a review of challenges, methods, and perspectives for non-intrusive load monitoring. Sensors **22**(15), 5872 (2022)
12. Liu, X., Zhang, F., Liu, H., Fan, H.: itimes: investigating semisupervised time series classification via irregular time sampling. IEEE Trans. Industr. Inf. **19**, 6930–6938 (2023)
13. Rao, D., Visin, F., Rusu, A.A., Teh, Y.W., Pascanu, R., Hadsell, R.: Continual unsupervised representation learning. In: Neural Information Processing Systems (2019)

14. Schwarzer, M., Anand, A., Goel, R., Hjelm, R.D., Courville, A., Bachman, P.: Data-efficient reinforcement learning with self-predictive representations. In: International Conference on Learning Representations (2020)
15. Wagy, M.D., Bongard, J.C., Bagrow, J.P., Hines, P.D.: Crowdsourcing predictors of residential electric energy usage. IEEE Syst. J. **12**, 3151–3160 (2017)
16. Xi, L., Yun, Z., Liu, H., Wang, R., Huang, X., Fan, H.: Semi-supervised time series classification model with self-supervised learning. Eng. Appl. Artif. Intell. **116**, 105331 (2022)
17. Yu, J., Jiang, J.: Learning sentence embeddings with auxiliary tasks for cross-domain sentiment classification. In: Conference on Empirical Methods in Natural Language Processing (2016)
18. Zhang, Yu., Yang, Q.: A survey on multi-task learning. IEEE Trans. Knowl. Data Eng. **34**, 5586–5609 (2022)

Handling Concept Drift in Non-stationary Bandit Through Predicting Future Rewards

Yun-Da Tsai[✉] and Shou-De Lin

National Taiwan University, Taipei, Taiwan
{f08946007,sdlin}@csie.ntu.edu.tw

Abstract. We present a study on the non-stationary stochastic multi-armed bandit (MAB) problem, which is relevant for addressing real-world challenges related to sequential decision-making. Our work involves a thorough analysis of state-of-the-art algorithms in dynamically changing environments. To address the limitations of existing methods, we propose the Concept Drift Adaptive Bandit (CDAB) framework, which aims to capture and predict potential future concept drift patterns in reward distribution, allowing for better adaptation in non-stationary environments. We conduct extensive numerical experiments to evaluate the effectiveness of the CDAB approach in comparison to both stationary and non-stationary state-of-the-art baselines. Our experiments involve testing on both artificial datasets and real-world data under different types of changing environments. The results show that the CDAB approach exhibits strong empirical performance, outperforming existing methods in all versions tested.

1 Introduction

Bandit algorithms have gained significant attention from both academic and industrial communities due to their applications in solving the exploration and exploitation dilemma in sequential decision-making problems. The classical multi-armed bandit (MAB) problem involves an agent selecting one out of K arms at each time step for a maximum of T rounds. Each arm is associated with an unknown reward distribution. At each time step, the agent selects one arm, and the actual reward drawn from the distribution is revealed to the agent. The objective is to maximize the cumulative expected reward over the T rounds. However, the agent must strike a balance between exploring the environment to find the optimal arm and exploiting the current best arm to maximize the cumulative reward. This trade-off is commonly referred to as the exploration and exploitation dilemma. In stationary cases, the reward distributions remain unchanged over time, whereas in non-stationary cases, the reward distributions can vary

Supplementary Information The online version contains supplementary material available at https://doi.org/10.1007/978-981-97-2650-9_13.

over time. Applications of bandit algorithms include clinical trials [36], information retrieval [11,23], strategic pricing [4], influence maximization in social networks [8], investment in innovation [3], packet routing and network optimization [2], on-line auctions [21], and on-line advertising [16,28,31].

The stochastic MAB problem assumes that rewards are generated independently from a stochastic distribution associated with each arm. However, in real-world scenarios, this assumption cannot hold true, as reward distributions continually evolve over time. For example, user interest in the content of a website, such as news, is likely to vary over time, as is the return on investment. To address these situations, the non-stationary MAB problem has recently attracted serious attention, where the reward distribution of each arm varies over time [19].

Over the years, several algorithms and real-world applications have been developed to study the non-stationary bandit problem. Most existing solutions start by defining different types of dynamic changing environments, each corresponding to a concept drift pattern that assumes how the reward distribution would evolve. For instance, the abruptly changing environment assumes that the reward distribution remains stationary on some time intervals until some abrupt change points, referred to as breakpoints. This leads to change point detection-based algorithms that restart the MAB bandit policy once the change point is detected. However, these algorithms can only react to the abruptly changing environment and not others. In the paper [9], the authors considered all non-stationary behavior as a concept drift problem and adopted the idea of active learning to adapt the learning model to cope with the changing distribution and react promptly. They further distinguished four different types of concept drift (Abrupt, Incremental, Gradual, Recurring) that may affect a stream of data [37]. The fact that concept drift can be distinguished into different types inspired us with a hypothesis that most of the reward distribution evolution often comes in distinguishable and predictable patterns. While the distribution of the product may remain the same, the conditional distribution of the product for that user evolves over time in similar patterns. One observation is that the dynamics of the changing rewards are not completely random. As illustrated in Fig. 1, which we collected from Google Trend top search terms and from the car traffic Loop Sensor dataset [10], the change in rewards (i.e., the number of clicks in Trend words or the amount of traffic) seems to follow a certain pattern rather than some unpredictable stochastic process. These concept drift patterns repeatedly appear in different users and products across different campaigns and ad-groups. Consequently, this paper argues that it is more robust to learn the drift in a data-driven manner, as opposed to relying on some predefined drift patterns. That is, we study how to learn and capture the characteristics of the changing environment through the observed sequences, and then let our MAB model make adjustments accordingly.

The key idea of this work is to automatically adapt to possible concept drift patterns in non-stationary environments. Previous works have primarily attempted to reduce the impact of concept drift by restarting the bandit algorithm after detecting drifts using ad-hoc test statistics [12,22,27] or by adjust-

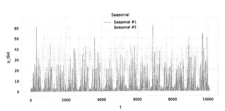
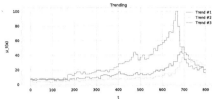

(a) Traffic flow that follows periodic events from the Loop Sensor dataset.

(b) User's interests that follows trending topics such as instant popular items from Google Trend top search terms.

Fig. 1. Concept Drift Patterns

ing the window size [15,32]. Instead of adapting the bandit policy based on past observations, we propose adapting the bandit policy using predicted future observations. In other words, we aim to recognize and predict the most probable future changes in the reward distribution so that the bandit policy can be adjusted proactively. Updating the reward estimator in the bandit policy beforehand using pseudo feedback is also a common approach when dealing with delayed feedback [20]. By incorporating the influence of temporal concept drift in reward distribution evolution, the bandit policy remains adaptive and maintains accurate mean reward estimation. Furthermore, this data-driven solution prevents us from making strong assumptions about predefined patterns for changing environments.

We propose the time-series Concept Drift Adaptive Bandit (CDAB) framework for dynamic Multi-Armed Bandits (MABs). CDAB utilizes time-series prediction to capture concept drift patterns from past observations, providing two advantages. First, CDAB can adapt to any type of changing environment. The advantage of a data-driven method is that it is not limited to predefined changing environments. Second, most previous works can only address non-stationary behavior by adapting from past observations. This forces the bandit policy to adjust only after the distribution change has occurred, leading to increased regret. CDAB employs look-ahead with future reward distribution prediction and performs pseudo updates, enabling the bandit policy to smoothly adapt to the most probable future observations. This approach shares a similar insight with delay feedback problems that utilize pseudo updates on the bandit policy before receiving actual feedback [20].

The contributions of this paper are threefold. First, we investigate different changing environments for non-stationary bandit problem and illustrate the concept drift patterns in real world scenarios. Second, we develop novel Time-series Concept Drift Adaptive Bandit (CDAB) framework. CDAB can be easily applied to any existing bandit policy and adapts to concept drift patterns in a data driven style and boost the performance by accurate pattern predictions. Third, we demonstrate the superior empirical results using both artificial and real world datasets.

2   Background and Related Works

The non-stationary multi-armed bandit (MAB) problem assumes that the reward distribution of each arm varies over time, which is typically the case in real-world scenarios. This non-stationary behaviour can also be referred as the concept drift problem [14], where the distribution of the data continually changes. Various types of dynamically changing environments have been studied. One class of non-stationary MABs has been investigated in an abruptly changing environment, where the mean rewards of any arm may switch abruptly at unknown times to unknown values. This setting is also known as piecewise-stationary. Another class of non-stationary MABs has been studied in a slowly-varying environment. The change in the mean reward between any two time step is small and is upper bounded by $\epsilon_T \in O(T^{-K})$, where $K \in \mathbb{R} > 0$ is known a constant. This indicates a local smoothness with a limited budget for temporal variation. There are other settings such as adversarial bandits and restful bandits.

A common approach for the piecewise-stationary environment is to apply change point detection with restarts. CD-UCB [25] proposes a change-detection (CD) based framework for multi-armed bandit problems under the piecewise-stationary setting. It actively detects change points and restarts the UCB indices using the cumulative sum (CUSUM) and Page-Hinkley Test (PHT). Adapt-EvE [18] relies on the UCBT algorithm incorporates a change-point detection test based on the Page-Hinkley statistics. Change-Point Thompson Sampling (CTS) [26] uses a Bayesian Change Point Detection. GLR-klUCB [45] uses a Bernoulli Generalized Likelihood Ratio Test in order to detect a change in the environment. AD-Switch [1], M-UCB [7] achieve (nearly) optimal mini-max regret bounds without knowing the number of changes.

Bandit algorithms in a slowly-varying environment often employ forget mechanisms. Sliding-window UCB (SW-UCB) [15] and sliding-window Thompson Sampling (SW-TS) [32] are adaptations that uses a sliding window approach for the non-stationary bandit problem. These methods uses a sliding window to estimate the mean rewards only from the most recent observations. ADWIN [6] is an adaptive sliding window method that automatically adapts the window size for change-point detection. Discounted Thompson Sampling (DTS) [29] and Discounted UCB [15] uses a discount method that gradually decays the weight for older observations as a forget mechanism. This maintains exploration of the environment and keeps track of the reward mean for each arm in recent observations.

3   Methodology

3.1   Problem Definition

We denote a stochastic bandit problem with changing reward distributions as follows: Let $A = \{1, \ldots, K\}$ be a set of arms. Let $t = 1, 2, \ldots, T$ denote the decision epochs. Let the mean reward associated with arm a at time t be $\mu_t(a) \in [0, 1]$. Let π_t be the current bandit policy at time t. At each time t, the agent pulls one of the K arms a_t at and receives reward $r_t \in [0,1]$, where r_t is a random variable with expectation $\mu_t = \mathbb{E}[r_t]$. The goal of the agent is to choose a sequence of arms at each time that minimizes the expected cumulative regret R:

$$R = \sum_{t=1}^{T} \mu_t(a_t^*) - \mathbb{E}[\sum_{t=1}^{T} \mu_t(a_t)] \tag{1}$$

Here a_t^* is the optimal arm with largest expected reward at time t. In stationary MAB problem, the gap between arms $\mu_t(a_t^*) - \mu_t(a_t)$, remains unchanged for all the time steps. However, in the problem with changing reward distributions, the gaps between arms vary throughout time. Since the mean rewards $\mu_t(a)$ are unknown to the agent, we further define a dynamic oracle [5] as the optimal policy to evaluate. The dynamic oracle optimizes the expected cumulative regret at each step $t \in T$ by always selecting the best arm a_t^*, with the highest expected reward. For simplicity, we will simply refer it as regret in the following.

3.2   Time-Series Concept Drift Adaptive Algorithm

The algorithm we proposed, called time-series concept drift adaptive bandit (CDAB), can be readily applied to any existing non-stationary bandit algorithms. The key idea is to predict potential future changes in the reward distribution and update the bandit policy in advance using look-ahead pseudo feedback, similar to methods that handle delayed [20]. The policy π_t is updated based on historical observations up to time t with selected arms $a_{1:t-1} \equiv (a_1, \cdots, a_{t-1})$ and received rewards $r_{1:t} \equiv (r_1, \cdots, r_{t-1})$, respectively. In addition to the bandit policy π, our proposed algorithm requires an additional time-series prediction model M. The reward prediction model M is updated with received rewards $r_{1:t} \equiv (r_1, \cdots, r_{t-1})$ and side information (context) $x_{1:t} \equiv (x_1, \cdots, x_{t-1})$ at each time step. Based on recent observations within a time window of length c, reward prediction model M will output the most possible empirical reward mean for each arm in future time step $r_{t+1}, r_{t+2} \cdots$. The predicted rewards $r_{t+1}, r_{t+2}, \cdots$ will be used to sample pseudo feedback from the future, allowing the policy π_t to be updated even before the actual rewards are received. In general, policy π_t make decision based on both past observations $r_{1:t}$ and predicted rewards $\hat{r}_{t:t+c}$. The detail algorithm is in Algorithm 1. In lines (3–4), we generate predicted pseudo rewards $\hat{r}_{t:t+c}$ from time-series prediction model M and acquire the pre-updated policy π_t'. In line (5–8), we select the arm based on policy π_t' and observe the reward at time t on line (9). In line (10–11), the pre-updated policy π_t' will be discard and we will update the policy π_t and time-series prediction model M with actual received rewards. This process will repeat from time step 1 to T.

4   Experiments

In this section, we present the experiments conducted on both synthetic and real-world data, involving several different dynamically changing environments.

Algorithm 1. Time-series Concept Drift Adaptive Algorithm

Require:
1: K: The number of arms
2: π: Bandit policy
3: π': Temporal Bandit policy updated with future rewards
4: c: Time window size for future reward prediction
5: M: Time-series prediction model
6: T: Total time steps
7:
8: Initialize π, M
9: **for** $t = 1, 2, \cdots$ to T **do**
10:     Generate pseudo rewards $\hat{r}_{t:t+c} = \arg\max M_t(r_{t-c:t})$.
11:     Update $\pi'_t \leftarrow \pi_t + \hat{r}_{t+c:t}$
12:     **for** $k = 1, 2, \cdots$ to K **do**
13:         $\hat{\mu}_t(a) = \pi'_t(a_k)$
14:     **end for**
15:     Select arm $\hat{a}^*_t = \arg\max_k \hat{\mu}_t(a)$.
16:     Observe actual reward r_t
17:     Update $\pi_{t+1} \leftarrow \pi_t + r_t$
18:     Update $M_{t+1} \leftarrow M_t + r_t$
19: **end for**

We compare the results with other state-of-the-art algorithms designed for non-stationary scenarios. We begin by illustrating the setup of our experimental campaign, followed by parameters tuning, dataset description, baselines used, and finally the presentation of the experiment results. In Appendix D we provide additional information on the running times of CDAB in comparison to the baseline algorithms. The detailed experiment setup is in Appendix A.

4.1   Dataset

Artificial Dataset. We generated synthetic data in various dynamically changing environments and investigated three specific types: abruptly-changing, slow-varying, and cyclic-occurring. The details for the generation of each dataset is in Appendix B.

Real-World Dataset. For real-world data, we utilized four public datasets OpenBandit [30], MovieLens [17], PM2.5 [24] and Loop Sensor [10]. While these datasets were initially designed for classification or regression task, we can transform them to better suit for bandit problem. This approach follows the methodology commonly used in the literature, as seen in [13,33–35]. Detailed information about these datasets is provided in Appendix C and Table 1a. Unlike artificial data, real-world data lacks solid and fixed patterns in reward distribution. However, our proposed method excels in capturing and learning arbitrary concept drift patterns.

4.2   Baseline

In the experiments, we compared CDAB with a set of state-of-the-art algorithms as baseline. The detail description of these algorithms are introduced in Sect. 2. These baselines are listed as following:

- Random: Random selects arms.
- TS: Classic Thompson Sampling algorithm that dawns random samples from Beta-Bernoulli posterior distribution.
- UCB: Classic Upper Confidence Bound algorithm
- D-TS, D-UCB: TS and UCB algorithm with a discount factor that decays the weights in past observations. Parameters γ controls the amount of discount.
- SW-TS, SW-UCB: TS and UCB algorithm with a time step sliding window for past observations. Parameter n controls the size of the sliding window.
- EXP3R: EXP3 variant with Drift Detection for non-stationary bandit.
- Ad-Switch: Breakpoints detection without knowing the number of changes for non-stationary bandit.
- CDAB-TS, CDAB-UCB: CDAB applied to TS and UCB algorithm.

The hyperparameters used in each baselines are listed in the Appendix Table 1b. The parameters listed were determined to be the best we found after several rounds of fine-tuning with manual hyperparameter selection. We focus on two parameters where γ is the discount factor and n is the sliding window size.

Table 1. Final cumulative regret at end of time step T. Mean and standard deviation of the cumulative regret for 20 repetition experiments is reported. CDAB showed strong empirical performance which outperforms all other baselines.

		Abruptly	Slowly	Cyclic	OpenBandit	Movielens	PM2.5	Loop Sensor
Stationary	Random	1217 ± 53	1318 ± 43	1258 ± 41	1325 ± 35	2143 ± 52	1389 ± 44	1489 ± 41
	TS	834 ± 42	772 ± 41	993 ± 45	1267 ± 34	1768 ± 49	1178 ± 46	1219 ± 33
	UCB	772 ± 24	710 ± 24	1029 ± 34	1215 ± 36	1441 ± 43	976 ± 41	1332 ± 36
Non Stationary	D-TS	389 ± 30	220 ± 19	356 ± 41	187 ± 19	377 ± 19	431 ± 31	269 ± 28
	D-UCB	361 ± 33	196 ± 20	378 ± 32	672 ± 24	881 ± 33	832 ± 38	778 ± 37
	SW-TS	307 ± 24	239 ± 23	331 ± 26	153 ± 27	729 ± 25	467 ± 29	436 ± 23
	SW-UCB	350 ± 27	207 ± 15	298 ± 19	271 ± 28	519 ± 31	541 ± 28	449 ± 21
	EXP3R	245 ± 22	331 ± 22	334 ± 26	349 ± 23	337 ± 32	497 ± 24	532 ± 36
	AdSwitch	197 ± 19	437 ± 32	371 ± 30	221 ± 35	342 ± 22	452 ± 29	311 ± 24
Ours	CDAB-TS	$\mathbf{60 \pm 13}$	$\mathbf{51 \pm 15}$	$\mathbf{53 \pm 10}$	$\mathbf{93 \pm 11}$	$\mathbf{134 \pm 29}$	$\mathbf{110 \pm 11}$	$\mathbf{88 \pm 14}$
	CDAB-UCB	$\mathbf{71 \pm 17}$	$\mathbf{67 \pm 17}$	$\mathbf{49 \pm 9}$	$\mathbf{89 \pm 18}$	$\mathbf{141 \pm 24}$	$\mathbf{99 \pm 21}$	$\mathbf{84 \pm 12}$

4.3   Experiment Results

The comparison of cumulative dynamic regrets of all datasets is summarized in Table 1, which is defined in Eq. 1. The cumulative dynamic regret over time t is shown in Fig. 2.

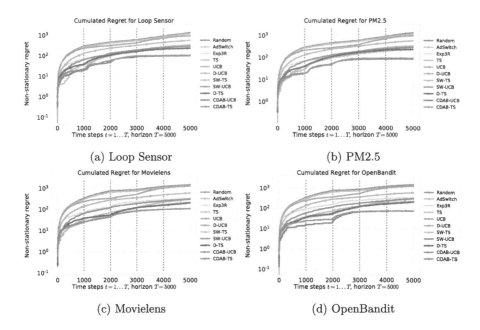

Fig. 2. Cumulative dynamic regret over time t for 4 real world dataset. CDAB showed strong empirical performance and appear to be best performing in all dataset. CDAB showed better decreasing rate in the beginning and converges to optimal more rapidly. With the help of CDAB, bandit policy can still selects optimal arm even with non-stationary rewards.

Artificial vs Real-World Dataset. First, we discuss the results of three artificial datasets that represent three different changing environments, namely abruptly-changing, slow-varying, and cyclic-occurring. The non-stationary baselines are designed to solve specific types of concept drifts and can perform well on at least one artificial dataset. However, they exhibit unstable performance on real-world datasets, where no algorithm can consistently perform well on all datasets. In contrast, our model is capable of learning the drifts present in the artificial datasets, which explains its high accuracy in time-series prediction and the significant advantages it offers to the bandit policy. Therefore, CDAB outperforms all other baselines by a considerable margin in all three changing environment datasets. When applied to real-world datasets, CDAB consistently outperforms other baselines by a decent margin.

Stationary vs Non-stationary. Standard algorithms for stationary environments, such as TS and UCB, fail to adapt to changing reward distributions, resulting in almost linear regret for most datasets. In contrast, non-stationary bandit algorithms yield much better performance in terms of dynamic regret, as the reward distributions in all datasets are highly variable. In general, our proposed approach, CDAB, outperforms all competitors significantly in terms of cumulative regret. Figure 2 illustrates that CDAB exhibits a better decreasing rate in the beginning and converges more rapidly, showing no further increase in regret. With the help of CDAB, the bandit policy can select the optimal arm even with non-stationary rewards, while the regret for other bandit policies still slightly increases. Although the prediction becomes less accurate in real-world scenarios, the concept drift adaptation in the bandit policy with time-series prediction still leads to improvements, to some extent. This implies that CDAB can enhance any bandit policy as long as the time-series prediction is effective to a certain extent.

Ablation Study. In this section, we aim to provide further insight and understanding into the performance advancements of CDAB through an ablation study. We will examine two key components that make up CDAB: (1) the look-ahead pseudo reward update and (2) the time-series prediction model M. The look-ahead pseudo reward update contributes to the adaptability of CDAB by enabling faster responses to distribution changes. On the other hand, the time-series prediction model M facilitates adaptation to various types of distribution changes and different rates of change. To investigate these components, we conduct ablation experiments using artificial dataset since the distribution changes observed in real-world datasets are not applicable for tracing purposes.

Table 2. The cumulative dynamic regret improves significantly when applying look-ahead for even only one step. *CDAB-TS-0* and *CDAB-UCB-0* are two variants that will no have pseudo rewards updates and thus have performance similar to regular TS, UCB algorithm. The bold font indicates the best performance among all.

	Abrubtly	Slowly	Cyclic			Abrubtly	Slowly	Cyclic
CDAB-TS-4	62 ± 13	53 ± 15	**52 ± 10**		CDAB-UCB-4	**70 ± 17**	**63 ± 17**	54 ± 9
CDAB-TS-3	**60 ± 13**	**51 ± 15**	53 ± 10		CDAB-UCB-3	71 ± 17	67 ± 17	**49 ± 9**
CDAB-TS-2	75 ± 15	61 ± 16	76 ± 17		CDAB-UCB-2	93 ± 19	73 ± 19	63 ± 13
CDAB-TS-1	87 ± 16	76 ± 17	88 ± 17		CDAB-UCB-1	104 ± 19	98 ± 19	77 ± 14
CDAB-TS-0	802 ± 40	760 ± 39	964 ± 37		CDAB-UCB-0	778 ± 21	723 ± 20	996 ± 29

In our first experiment, we emphasize the significance of the look-ahead pseudo reward update. We compare the performance of CDAB with different look-ahead time windows, denoted as c. Table 2 illustrates that even a single-step look-ahead significantly improves the cumulative dynamic regret. This finding is particularly noteworthy considering that our prediction model M achieves near-perfect predictions on the artificial dataset.

In the second experiment, we would like to highlight that the prediction model is directly fitted to the observed data, allowing it to adapt to arbitrary changing environments or changing rates. Most related works often rely on a hyperparameter to determine a fixed adaptive rate for the bandit policy, such as decay rate or sliding window size. To address this limitation, we constructed four challenging artificial datasets. For each changing environment (abrupt, slow,

Table 3. (a) The increase percentage of cumulative dynamic regret compared to the results in Table 1. The trajectory is 3 times longer so the baseline of the increase percentage is 3. The regret for CDAB grows about 4× while others grows about 5–7× more. The bold font indicates the best performance among all. (b) The influence of the quality of time-series prediction M. Linear autoregressive model failed to fit non-linear distribution and thus cause the improvement degrade significantly. The random prediction shows that CDAB will completely fail in this case. The bold font indicates the best performance among all.

		Abrubtly*	Slowly*	Cyclic*	Hybrid
Baseline	D-TS	5.39	5.42	5.71	7.13
	D-UCB	5.97	6.11	5.71	6.44
	SW-TS	5.62	5.87	5.91	6.77
	SW-UCB	6.12	6.01	5.91	6.76
	EXP3R	5.75	5.62	5.85	6.24
	AdSwitch	5.12	5.23	5.76	6.12
Ours	CDAB-TS	**3.86**	**4.34**	4.13	**4.28**
	CDAB-UCB	4.23	4.41	**4.09**	4.32

	Abrubtly	Slowly	Cyclic
CDAB-TS	**60 ± 13**	**51 ± 15**	53 ± 10
CDAB-UCB	71 ± 17	67 ± 17	49 ± 9
AR-TS	613 ± 32	581 ± 31	709 ± 35
AR-UCB	674 ± 34	589 ± 34	671 ± 34
Random-TS	812 ± 51	750 ± 44	835 ± 41
Random-UCB	829 ± 49	781 ± 46	871 ± 47

cyclic), we generated three datasets with different parameters controlling the change rates and concatenated all datasets into one long trajectory. Additionally, we created a challenging dataset by concatenating different changing environments.

Table 3a presents the increase percentage of cumulative dynamic regret compared to the results shown in Table 1. Since the trajectory is three times longer, the baseline increase percentage is three. The regret for CDAB grows approximately four times, while the other models experience a growth of five to seven times more.

In the third experiment, we compared the influence of the quality of M. When the data is completely unpredictable or when M is too weak to make accurate predictions, the assumptions of CDAB may not hold. As a result, the worst-case scenario of CDAB can occur, where no improvement gain is observed. In Table 3b, we compared three models: LSTM, linear autoregressive, and Random. The linear autoregressive model failed to fit the non-linear distribution, leading to a significant degradation in improvement. The random prediction clearly demonstrates that CDAB will completely fail in such cases.

5   Conclusion

This research paper proposes a novel framework, called CDAB, which can be integrated with existing bandit solutions to effectively address non-stationary bandit problems. By combining time-series prediction and bandit policies, CDAB tackles the challenge of concept drift. The experiments conducted demonstrate a significant improvement in efficiency, with a competitive performance in terms of rewards obtained. Furthermore, ablation studies are conducted to examine each component of CDAB, highlighting the improvements it brings to the bandit policy. In addition to the empirical analysis, future work will involve a theoretical analysis of CDAB's finite-time regret. Furthermore, the application of CDAB to other algorithms will be explored.

References

1. Auer, P., Gajane, P., Ortner, R.: Adaptively tracking the best bandit arm with an unknown number of distribution changes. In: Conference on Learning Theory, pp. 138–158. PMLR (2019)
2. Awerbuch, B., Kleinberg, R.D.: Adaptive routing with end-to-end feedback: distributed learning and geometric approaches. In: Proceedings of the Thirty-Sixth Annual ACM Symposium on Theory of Computing, pp. 45–53 (2004)
3. Bergemann, D., Hege, U.: The financing of innovation: learning and stopping. RAND J. Econ. **36**(4), 719–752 (2005)
4. Bergemann, D., Välimäki, J.: Learning and strategic pricing. Econometrica: J. Econometric Soc. **64**(5), 1125–1149 (1996)
5. Besbes, O., Gur, Y., Zeevi, A.: Stochastic multi-armed-bandit problem with non-stationary rewards. In: Advances in Neural Information Processing Systems, vol. 27 (2014)
6. Bifet, A., Gavalda, R.: Learning from time-changing data with adaptive windowing. In: Proceedings of the 2007 SIAM International Conference on Data Mining, pp. 443–448. SIAM (2007)
7. Cao, Y., Wen, Z., Kveton, B., Xie, Y.: Nearly optimal adaptive procedure with change detection for piecewise-stationary bandit. In: The 22nd International Conference on Artificial Intelligence and Statistics, pp. 418–427. PMLR (2019)
8. Carpentier, A., Valko, M.: Revealing graph bandits for maximizing local influence. In: Artificial Intelligence and Statistics, pp. 10–18. PMLR (2016)
9. Cavenaghi, E., Sottocornola, G., Stella, F., Zanker, M.: Non stationary multi-armed bandit: empirical evaluation of a new concept drift-aware algorithm. Entropy **23**(3), 380 (2021)
10. Chen, C., Petty, K., Skabardonis, A., Varaiya, P., Jia, Z.: Freeway performance measurement system: mining loop detector data. Transp. Res. Rec. **1748**(1), 96–102 (2001)

11. Combes, R., Magureanu, S., Proutiere, A., Laroche, C.: Learning to rank: regret lower bounds and efficient algorithms. In: Proceedings of the 2015 ACM SIG-METRICS International Conference on Measurement and Modeling of Computer Systems, pp. 231–244 (2015)
12. Dries, A., Rückert, U.: Adaptive concept drift detection. Stat. Anal. Data Min. ASA Data Sci. J. **2**(5–6), 311–327 (2009)
13. Dudík, M., Langford, J., Li, L.: Doubly robust policy evaluation and learning. arXiv preprint arXiv:1103.4601 (2011)
14. Gama, J., Žliobaitė, I., Bifet, A., Pechenizkiy, M., Bouchachia, A.: A survey on concept drift adaptation. ACM Comput. Surv. (CSUR) **46**(4), 1–37 (2014)
15. Garivier, A., Moulines, E.: On upper-confidence bound policies for non-stationary bandit problems. arXiv preprint arXiv:0805.3415 (2008)
16. Guo, D., et al.: Deep Bayesian bandits: exploring in online personalized recommendations. In: Fourteenth ACM Conference on Recommender Systems, pp. 456–461 (2020)
17. Harper, F.M., Konstan, J.A.: The movielens datasets: history and context. ACM Trans. Interact. Intell. Syst. (TIIS) **5**(4), 1–19 (2015)
18. Hartland, C., Gelly, S., Baskiotis, N., Teytaud, O., Sebag, M.: Multi-armed bandit, dynamic environments and meta-bandits (2006)
19. Hernandez-Leal, P., Kaisers, M., Baarslag, T., de Cote, E.M.: A survey of learning in multiagent environments: dealing with non-stationarity. arXiv preprint arXiv:1707.09183 (2017)
20. Huang, K.H., Lin, H.T.: Linear upper confidence bound algorithm for contextual bandit problem with piled rewards. In: Bailey, J., Khan, L., Washio, T., Dobbie, G., Huang, J., Wang, R. (eds.) PAKDD 2016. LNCS, vol. 9652, pp. 143–155. Springer, Cham (2016). https://doi.org/10.1007/978-3-319-31750-2_12
21. Kleinberg, R., Leighton, T.: The value of knowing a demand curve: Bounds on regret for online posted-price auctions. In: 44th Annual IEEE Symposium on Foundations of Computer Science, Proceedings, pp. 594–605. IEEE (2003)
22. Klinkenberg, R., Joachims, T.: Detecting concept drift with support vector machines. In: ICML, pp. 487–494 (2000)
23. Kveton, B., Szepesvari, C., Wen, Z., Ashkan, A.: Cascading bandits: learning to rank in the cascade model. In: International Conference on Machine Learning, pp. 767–776. PMLR (2015)
24. Liang, X., Li, S., Zhang, S., Huang, H., Chen, S.X.: PM$_{2.5}$ data reliability, consistency, and air quality assessment in five Chinese cities. J. Geophys. Res. Atmos. **121**(17), 10–220 (2016)
25. Liu, F., Lee, J., Shroff, N.: A change-detection based framework for piecewise-stationary multi-armed bandit problem. In: Proceedings of the AAAI Conference on Artificial Intelligence, vol. 32 (2018)
26. Mellor, J., Shapiro, J.: Thompson sampling in switching environments with Bayesian online change detection. In: Artificial Intelligence and Statistics, pp. 442–450. PMLR (2013)
27. Nishida, K., Yamauchi, K.: Detecting concept drift using statistical testing. In: Corruble, V., Takeda, M., Suzuki, E. (eds.) DS 2007. LNCS, vol. 4755, pp. 264–269. Springer, Heidelberg (2007). https://doi.org/10.1007/978-3-540-75488-6_27
28. Pandey, A., Singh, P., Iyengar, L.: Bacterial decolorization and degradation of azo dyes. Int. Biodeterior. Biodegradation **59**(2), 73–84 (2007)
29. Raj, V., Kalyani, S.: Taming non-stationary bandits: a Bayesian approach. arXiv preprint arXiv:1707.09727 (2017)

30. Saito, Y., Aihara, S.: Large-scale open dataset, pipeline, and benchmark for bandit algorithms. arXiv preprint arXiv:2008.07146 (2020)

31. Tóth, B., Sachidanandan, S., Jørgensen, E.S.: Balancing relevance and discovery to inspire customers in the IKEA app. In: Fourteenth ACM Conference on Recommender Systems, pp. 563–563 (2020)

32. Trovo, F., Paladino, S., Restelli, M., Gatti, N.: Sliding-window thompson sampling for non-stationary settings. J. Artif. Intell. Res. **68**, 311–364 (2020)

33. Tsai, T.H., Tsai, Y.D., Lin, S.D.: lil'HDoC: an algorithm for good arm identification under small threshold gap. arXiv preprint arXiv:2401.15879 (2024)

34. Tsai, Y.D., Lin, S.D., Lin, S.D.: Fast online inference for nonlinear contextual bandit based on generative adversarial network. arXiv preprint arXiv:2202.08867 (2022)

35. Tsai, Y.D., Tsai, T.H., Lin, S.D.: Differential good arm identification. arXiv preprint arXiv:2303.07154 (2023)

36. Zelen, M.: Play the winner rule and the controlled clinical trial. J. Am. Stat. Assoc. **64**(325), 131–146 (1969)

37. Žliobaitė, I.: Learning under concept drift: an overview. arXiv preprint arXiv:1010.4784 (2010)

Author Index

Z. Wang and C. W. Tan (Eds.): PAKDD 2024 Workshops, LNAI 14658, p. 175, 2024.
https://doi.org/10.1007/978-981-97-2650-9

Printed in the United States
by Baker & Taylor Publisher Services